SYNCHRONOUS MANUFACTURING

Principles for World Class Excellence

DR. M. MICHAEL UMBLE
Associate Professor of Operations Management
Baylor University
Waco, Texas

DR. M. L. SRIKANTH, PRESIDENT
The Spectrum Management Group, Inc.
New Haven, Connecticut

GF61AA
PUBLISHED BY
SOUTH-WESTERN PUBLISHING CO.
CINCINNATI, OH   WEST CHICAGO, IL   DALLAS, TX   LIVERMORE, CA

Library of Congress Cataloging-in-Publication Data

Umble, M. Michael
  Synchronous manufacturing.

  (APICS South-Western series in production and operations management)
  Bibliography: p.
  Includes index.
  1. Production management. 2. Production control.
I. Srikanth, Mokshagundam L.   II. Title.
III. Series.
TS155.U48 1990   658.5   89-90902
ISBN 0-538-80493-9

1 2 3 4 5 6 7 8 9 M 7 6 5 4 3 2 1 0 9

Printed in the United States of America

CONTENTS

This book is dedicated to our parents,
Gerald Wesley and Gertrude Winona Umble
and Sharada and M. V. L. Rao,
for their love and guidance.

FOREWORD

Any foreword is a blend of introduction, summary, and personal opinion. While this opinion should convey the importance of the book, its main purpose is to suggest how the book should be used. In this sense, objectivity is the yardstick to which we should adhere. But how can one be objective about a book that details, explains, interprets, and further develops one's own brainchild?

While reading the manuscript, I found myself reacting in two ways. I felt a sense of pride and took great satisfaction in what was so clearly and accurately described. At other times, I felt a sense of indignation at what initially appeared to be distortion. Upon a second reading, this reaction was dispelled by what proved to be plausible explanations that were at times real eye-openers. In spite of my efforts, maintaining total objectivity in the preparation of this foreword has been impossible, and it should be read with that in mind.

I concluded that in order to keep this book in the proper perspective, the best foreword should describe the background of the book's title. To the best of my knowledge, the phrase *synchronous manufacturing* was first coined in the summer of 1984. At that time, the first stage of implementing the ideas detailed in this book had already been completed at Saginaw Steering Gear, a division of General Motors. Blair Thompson, then head of the division (today an executive vice president of GM), asked me to reconsider the name optimized production technology under whose banner these considerable efforts had been achieved. He claimed (and rightfully so) that the software was the least important component of this effort and that the principles, or "thoughtware," as they were called, were the most powerful part. Thus, it was not particularly helpful to use a name primarily associated with a commercial software package.

It was not the first time I had considered changing the name. As a matter of fact, it had already been changed once before. The original name, established in 1979, was optimized production timetables. The main thrust was computerized scheduling, and the thoughtware was in its infancy. During the following years, the fundamental understanding grew by such leaps and bounds that in 1982, the name (though not the initials) was officially changed to optimized production technology.

When Blair suggested that I come up with a new name, I was already convinced of the need to change it again. I had three problems with the existing name. The first problem was the word *optimized*. This word implies that there is some imaginary maximum level of performance that we should strive to achieve. This notion is in direct conflict with the concept of "a process of ongoing improvement,"—a concept that began to appear even then. The second problem was the use of the word *production*. The thoughtware had already been developed enough to confirm that its ramifications went far beyond the scope of production. The third problem was the use of the word *technology*, which again was too narrow and restrictive. My conviction that the term *optimized production technology* was totally inappropriate ran so deep that in my book *The Goal*, published in 1984, the word does not appear.

When Blair suggested the term *synchronous manufacturing*, I supported it. This name was chosen to reflect in a positive way what the motto of the thoughtware had formulated in a negative way: "The total of local optima is not the optimum of the total." Moreover, it enabled us to put under the same umbrella all the procedures and concepts of JIT and TQC that were filtering into the United States at that time. For the same reason, the word *production* was replaced by the word *manufacturing* to broaden the scope. Since then, this phrase has been accepted to such a degree at GM that a synchronous manufacturing manager has been named in every division of the components group in order to supervise all those improvements activities.

As much as I supported the name *synchronous manufacturing*, I never actually used it. At the rate that the knowledge was evolving and with the growing awareness that psychology played a larger role than previously thought, I felt that even this new phrase was still too restrictive.

At the beginning of 1987, I adopted the current name—theory of constraints. A better understanding of the psychology caused me to make a shift from emphasizing rules/principles to a focusing–iterative process. Moreover, the significant ramifications that this process has for such areas as accounting, distribution, marketing, and product design almost forced that choice of words. Further, by 1988, the rapidly evolving understanding exposed that the main constraints in organizations were not physical constraints (capacity, market, vendors) but policy constraints. This last development almost pushed aside the issue of synchronization and thrust the need for processes to identify and elevate these devastating, constraining policies onto center stage. What more will evolve in the future? Only time will tell.

As with many other fields of science, this rapid rate of development is not a revolutionary process but an evolutionary process. New developments did not make obsolete nor did they contradict the previous steps. Rather, they opened new frontiers. New developments did not make the previous developments less valid; on the contrary, they turned them into springboards for more generic and powerful steps.

Over time, the need and thus the pressure was building for books to cover these developments. It was necessary to draw a line at some point in time and capture the know-how in a form that could be used to inform as well as instruct others. Some not-too-successful attempts were made previously. *The Goal*, as powerful as it is in delivering the message, was more suitable for general reading than for instructional use. Srikanth and Cavallaro's attempt to supplement it with *Regaining Competitiveness* turned out to be more successful in plant environments than in satisfying university course needs. *The Race*, coauthored with Robert Fox, was another attempt to meet both objectives. It succeeded as a reference book, but it was too condensed to be used as a textbook. My opinion is that this book is the first that can really do the job of informing production and operations managers as well as instructing students in related programs.

I would caution the reader that this book is not intended to present a complete and final picture. Rather, it captures a moment in time of something that is continually evolving. Instructors should be careful not to present it otherwise. We have enough sacred cows now; we should not add more. I recommend that as the concepts and principles are taught, students be encouraged to scrutinize them critically. Moreover, instructors should add additional questions to the excellent ones that follow each chapter—questions that force the student to realize that any valid approach which answers recognized problems will lead to many more new questions.

The following provide examples of this type of questioning:

Chapter 2: How can we numerically quantify the impact of competitive edge factors by using the operational measurements?

Chapter 3: If we should not use line balancing methods, how should a line be designed?

Chapter 4: Accepting the idea of CCRs and non-CCRs, what are suitable numerical local performance measurements for our resources?

Chapter 5: How can we determine the size of the process batches and transfer batches in production, engineering, distribution, and service environments?

Chapter 6: If the buffer is expressed in time, can we talk in terms of its location? What are the ramifications?

Chapter 7: Should we have only one type of time buffer? Should we have other types of buffers?

Chapter 8: What will A-, V-, and T-plants evolve into if the concepts in the previous chapters are fully implemented?

I am convinced that if used properly, this book will contribute significantly to the know-how of every manager or student who intends to be involved with any type of organization. Mike and Sri have accomplished the very demanding and most difficult job of molding into an excellent textbook that which was available only in scattered publications and as unwritten material. I congratulate them for their accomplishment.

Dr. Eliyahu M. Goldratt

PREFACE

There is a silver lining in the dark cloud that hangs over American manufacturing companies. A small but growing number of companies have demonstrated the ability to compete successfully in the international marketplace. A handful have done so in spectacular fashion, and their stories have become the staples of the conference and seminar circuits. The majority of successes have been achieved slowly and painstakingly but are just as impressive. A few from our own direct experience are listed below:

Textile Goods Firm:
Improved inventory turns from 20 to 80
Reduced lead times from 16 weeks to 7 weeks
Reduced finished goods inventory by $60 million
Improved ability to react to market opportunities quickly, which in turn led to export opportunities

Electronic and Electrical Products Firm:
Improved delivery performance from 70 percent to over 90 percent
Reduced past due orders by 50 percent
Reduced work-in-process inventory by over 30 percent
Improved synergy and communications between different groups

Contract Furniture Firm:
Reduced production lead times from 12 to 4 weeks
Reduced total inventory by 50 percent
Reduced quality problems by 35 percent
Opened up new marketing opportunities
Expanded into new product line requiring new equipment without need for additional buildings

Automotive Components Firm:
Improved inventory turns from 15 to over 50
Improved on-time deliveries from 70 percent to over 90 percent
Eliminated the use of overtime

Eliminated the hockey stick phenomenon at the end of the month
Increased total employee involvement

Why is it when almost every company in America sorely needs results similar to these, there are so few success stories? We believe that this is due to one simple and basic fact:

Widely used management policies and practices that are based on the standard cost system are not applicable in today's competitive environment.

These policies and practices evolved during a period of time when demand exceeded capacity and American firms dominated every industrial segment. Today, the marketplace is vastly different. Competition is global and more intense; markets are finely segmented and more demanding. The realization that the management infrastructure based on standard cost principles is at the very heart of our competitive problems is now well recognized (reference any recent conference of the NAA, or the works of Skinner, Hays, and Goldratt). Along with this realization comes a new series of critical questions such as the following:

1. Why does the standard cost system, which served so well, now fail us?
2. What are the new principles and strategies for managing manufacturing operations?

Synchronous manufacturing provides the answers to these and many other critical questions. This book develops the fundamental concepts and techniques of synchronous manufacturing. It is written with both the practitioner and the student of operations management in mind.

After reading this book, the practitioner will be better able to identify those managerial policies and practices that are the cause of a company's competitive problems. Moreover, the practitioner should be able to begin applying the synchronous manufacturing principles to improve overall company performance.

By studying this book, students will obtain a thorough understanding of the fundamental issues at play in the management of manufacturing operations. Students will also learn proven and systematic methods of analyzing and improving manufacturing operations.

This book is the first comprehensive work to fully explain the continually evolving principles of synchronous manufacturing. This book is the first of two volumes to be published. The second volume will be titled *Synchronous Manufacturing: Implementation Strategies and Case Studies*. The second volume will address the potential pitfalls of implementing major changes in any organization and will fully illustrate how such a process of change must be carefully managed. The second volume will also develop recommended implementation strategies in detail, and present three in-depth case studies.

Our efforts in undertaking this project will be more than amply rewarded if our work plays a role, however small, in restoring U.S. manufacturing to a position of world class excellence.

The authors would like to acknowledge several individuals who have contributed significantly to this text. Foremost is Eli Goldratt. His dedication, vision, and leadership in expanding the existing body of knowledge have made this book possible. Tim Fry of the University of South Carolina, Byron Finch of Miami University, Boaz Ronen of New York University, Martin Edelman, and Ronald Haft all provided careful and thought-provoking reviews of the manuscript. Elisabeth Umble spent many long hours editing the manuscript. And our sincerest thanks to Jim Cox of the University of Georgia who guided the development of the text as consulting editor and provided detailed and insightful critiques of each draft of the manuscript.

M. Michael Umble
Mokshagundam L. Srikanth

1

The Manufacturing Environment Today

WALLS OF PROTECTION

For nearly four centuries, from 404 B.C. to 37 B.C., Sparta was the most powerful city-state of ancient Greece. The Spartans were known far and wide for their disciplined and well-trained armies. During this period in history, it was common practice to build walls around cities to protect against hordes of bandits and invading armies. But Sparta had no such wall as protection. When asked why they did not build a wall, the Spartans replied by saying "The shields of our soldiers are our wall; they are the best protection against our enemies."

How effective were physical walls in protecting the ancient cities? Not very. Walls are a very inflexible and limited form of protection, being only defensive in nature. Many cities with seemingly impenetrable walls fell because a stronger enemy simply surrounded and blockaded the city until the inhabitants were starved out.

Wall building has not become a lost art. The instinct to defend territory, making it secure against evil forces and uncertainty, runs deep in our society. High walls usually evoke a strong sense of security. However, history has repeatedly taught us that such walls may only provide a false sense of security. And when people rely on walls for protection instead of their own strength and resilience, they become highly vulnerable.

In this country, we have built many "walls" for protection in our manufacturing plants. These "walls," both physical and nonphysical in nature, are intended to insulate individual departments and functional areas from uncertainty and the numerous disturbances that regularly occur within the manufacturing environment. They seemingly give each department or functional area a greater sense of independence, importance, and security. But in reality, the "walls" only act to divide the plant into small empires, weakening the total system.

Perhaps the most visible "walls" in our plants are the near mountainous "walls of inventory." Our plants often contain staggering amounts of finished goods, component stores, work in process, and raw materials. These inventories are meant to provide protection against manufacturing environment uncertainties such as variable customer demand, process and setup variability, unreliable vendors, quality problems, and poor schedules. It will be demonstrated in later chapters that these "walls of inventory" actually prevent our manufacturing plants from becoming competitive.

It can be said that "Every evil in a manufacturing environment eventually manifests itself in the form of additional inventory." In other words, inventory has become a mechanism for covering up and ignoring problems to the exclusion of finding solutions to those problems. Excessive inventory is absolutely dysfunctional to the efficient operation of the manufacturing plant!

The inherent flaw of the wall-building mentality in ancient times was that the inhabitants of the cities usually became overly reliant upon their walls for protection. As a result, the training of their soldiers was neglected, hastening the day when their city would fall. Unfortunately, the same mentality exists in many of our manufacturing plants today. Managers depend on a variety of artificial "walls" to protect their operations, and the results are less than satisfactory. Manufacturing managers must learn to protect their plants in the same way that the Spartans protected their city—without reliance upon walls.

This book will provide a thorough analysis of the "walls" that exist in our manufacturing plants and the assumptions and misguided policies that provide their foundation. The dysfunctional "walls" in our organizations will be exposed. We will learn how to demolish or circumvent these dysfunctional "walls," develop effective managerial and logistical procedures, and rely upon the abilities of our "soldiers" to make our plants competitive. In the process, an understanding of how to analyze a manufacturing plant and implement a process of ongoing improvement within the organization will be developed.

THE BIG PICTURE

The United States is the foremost industrial nation in the world today. The United States possesses the highest gross national product (GNP) and the most attractive markets in the world by a substantial margin. Historically, the traditional industrial sector in the United States has been the leader in productivity growth, the wellspring of innovation, and the primary stimulus of a rising standard of living. The more recent spectacular growth in the domestic service sector was possible only because of the existence of the strong industrial base. Productivity gains in manufacturing have provided the necessary support for the increasing demand for services.

But there is increasing concern about the future of manufacturing in this country. The era of unchallenged U.S. leadership in manufacturing innovation, process engineering, productivity, and market share is long gone. It is clear that the United States has lost its predominant position in a number of key industries such as steel, automobiles, textiles, and electronics. And it is apparent that competitive pressures continue to mount in numerous other key industries. Foreign manufacturers offering high-quality, low-priced products continue to penetrate scores of once secure domestic markets.

The challenges of global competition and rapidly changing technology require an increased managerial understanding of the total manufacturing process. Although some firms have managed to improve their operations, the typical U.S. response to these challenges has been disappointing. The most significant problem in domestic manufacturing firms today is the lack of commitment to the necessary organizational, managerial, and logistical changes that must occur in order to become or remain competitive. Manufacturing must be recognized as an integrated system requiring coordination and cooperation between functional areas if our plants are to meet the competitive challenge.

THE CHALLENGE TO MANUFACTURING

Losing the Competitive Edge

In order to fully understand and appreciate the current competitive environment, consider some of the changes that have occurred over the last 25 years. One clear indication of change is that even though manufacturing productivity in this country remains the highest in the world, it has been virtually equaled in recent years by Japan, West Germany, and France. Continuing productivity trends indicate that these countries are poised to surpass us. The data in Table 1.1 indicate that the growth in overall productivity (as measured by output per worker) in the United States has been among the lowest for industrial nations since 1977. Moreover, according to a 1987 *Wall Street Journal* article, defense production reportedly accounted for more than 40 percent of U.S. productivity increases since 1979. And nondefense productivity is estimated to have risen less than 1 percent annually since 1979. [21] Table 1.2 shows the annual percent changes in nonfarm productivity growth for the United States from 1980 through 1987.

Another indicator of the significant changes that are occurring on a global basis is the balance of trade account. The balance of trade account is a telling yardstick of a nation's ability to compete in the world marketplace and to maintain long-term economic superiority. As illustrated in Table 1.3, published government figures show that the U.S. trade imbalance has reached

TABLE 1.1   OUTPUT PER WORKER IN 1987 AND AVERAGE ANNUAL PERCENT GROWTH SINCE 1977, SELECTED COUNTRIES

COUNTRY	1987 OUTPUT PER WORKER	PERCENT GROWTH SINCE 1977
United States	$38,896	7 %
Canada	$37,150	10 %
France	$33,171	23 %
West Germany	$31,546	18 %
Japan	$27,508	35 %
Korea	$13,280	63 %

Source: Reprinted with permission of *Wall Street Journal*, © Dow Jones & Company, Inc. (1988). All Rights Reserved Worldwide.

TABLE 1.2   OUTPUT PER MAN-HOUR IN U.S. NON-FARM BUSINESS, ANNUAL PERCENT CHANGE, 1980 -1987

YEAR	PRODUCTIVITY CHANGE
1980	-0.4
1981	1.0
1982	-0.6
1983	3.3
1984	2.1
1985	1.2
1986	1.6
1987	0.8

Source: Economic Indicators, 100th Congress, June 1988

catastrophic proportions. The United States last recorded a trade surplus in 1981. In 1982, the trade deficit was a relatively modest $8.7 billion. But since 1982, the deficit has grown at an average annual rate of $29 billion per year to a record high of $154 billion in 1987. [22]

The huge role of manufactured imports and exports in the overall trade deficit is easy to see. Table 1.3 also shows the manufacturing trade balance for the United States from 1980 to 1987. The data clearly indicate that the

**TABLE 1.3   U.S. BALANCE OF TRADE ACCOUNT,
IN BILLIONS OF DOLLARS, 1980 - 1987**

YEAR	TOTAL SURPLUS(+) OR DEFICIT (-)	MANUFACTURING SURPLUS(+) OR DEFICIT (-)
1980	+1.9	+22.0
1981	+6.9	+15.5
1982	-8.7	-2.7
1983	-46.2	-29.9
1984	-107.1	-78.1
1985	-115.1	-101.4
1986	-138.8	-128.7
1987	-154.0	-137.7

Source: Economic Indicators, 100th Congress, June 1988; and U.S. Department of
Commerce, International Trade Administration, June 1988

competitive imbalance in manufacturing is the primary culprit for the
continuing United States trade deficit. The long-term problem is that the United
States has become a net consumer nation, which implies that less capital
is available for investment purposes both at home and abroad. As foreign
firms continue to invest their gains in the United States, increasingly larger
parts of the domestic economy are becoming foreign-owned. The obvious
implication is that the United States is receiving a smaller percentage of the
return from our productive efforts as the return on investment in the foreign-
owned corporations escapes overseas. It is becoming increasingly clear that
in order to maintain a high standard of living, the United States must become
more competitive in manufacturing.

A special report in the March 3, 1986, issue of *Business Week* described
in great detail an increasingly popular corporate strategy that is occurring
in the United States and is resulting in what is termed the "hollow corporation."
This phenomenon is described as follows:

> In industry after industry, manufacturers are closing up shop or curtailing
> their operations and becoming marketing organizations for other
> producers, mostly foreign. . . . The result is the evolution of a new kind
> of company: manufacturers that do little or no manufacturing and are
> increasingly becoming service-oriented. They may perform a host of profit-
> making functions—from design to distribution—but lack their own
> production base. In contrast to traditional manufacturers, they are hollow
> corporations. [23]

The hollowing phenomenon is another symptom of the inability of
American firms using traditional and commonly accepted manufacturing

management practices to compete with foreign competitors. The simple loss of blue collar jobs to low-wage countries, which first occurred in the 1960s and 1970s, was unfortunate. But now U.S. firms are also exporting technology, management functions, and even the design and engineering skills that are crucial to innovation. By shifting manufacturing operations to low-cost sites overseas or shopping in foreign markets for parts and components, U.S. corporations are chiseling away at their own critical mass necessary for a strong industrial base.

Some firms view the hollowing process as a way to remain competitive in the short run. But in the long run, it is likely to turn into a disaster for the U.S. economy. The proposition that the United States can retain its position of economic leadership without a strong industrial base appears to be a dangerous myth. Many experts agree that the United States cannot continue to prosper as an increasingly service-based economy. Robert A. Lutz, chairman of Ford of Europe, Inc., proclaims: "You're seeing a substantial deindustrialization of the U.S., and I can't imagine any country maintaining its position in the world without an industrial base." Japanese industrialist Tsutomu Ohsima, who serves as senior managing director of Toyota Motor Corp., observes: "You can't survive with just a service industry." And Akio Morita, chairman of Sony Corp., puts it even more strongly with the warning that unless the United States shores up its manufacturing base, "it could lose everything." [23] It is absolutely essential that U.S. firms rediscover their manufacturing management expertise and fight to keep their competitive advantage.

Regaining the Competitive Edge

How did the United States lose its competitive edge in manufacturing, and how can it be regained? Many reasons have been offered to try to explain the inability of U.S. industry to compete with foreign competitors. Among the reasons most often heard are the high cost of domestic labor, inflexible union work rules, outdated factories and equipment, and a wide range of favorable government policies in competing nations. However, under close scrutiny, these arguments can all largely be discounted. For example, a detailed survey of 171 of the more than 1,000 plant closings during the 1970s, published in the *Harvard Business Review*, yielded a number of interesting results. The average age of the factories at closing was 19.3 years, the median age was 15 years, and a third of the plants were no more than 6 years old. Furthermore, while two-thirds of the closed plants were unionized, managers in most of those plants characterized labor's attitude as tolerable or better. Fewer than 20 percent of the closed plants were plagued by militant, uncompromising attitudes or by a history of work stoppages. [24]

Examination of a cross section of U.S. industry reveals that a number of domestic companies have been able to compete quite successfully while others in the same industry under similar conditions have gone bankrupt. Why do some U.S. companies fail and others succeed? The difference is not automated production systems, robots, cheap labor, or a superior work ethic. Care must be taken to prevent these and other largely insignificant variables from muddying the water and obscuring the more fundamental and most significant explanation. The primary difference is management!

An analysis of successful Japanese companies typically reveals a superior manufacturing management system. If U.S. industry is to regain its competitive edge in manufacturing, the management of our manufacturing logistical systems must be significantly improved. Management must be willing to question the rationale for the "walls" that exist in our organizations. They must challenge the assumptions underlying our traditional methods of manufacturing planning and control. There can be no other way. The quest for understanding begins by briefly tracing the development of manufacturing logistical systems in this country.

DEVELOPMENT OF LOGISTICAL SYSTEMS

During the first 20 to 25 years after World War II, U.S. industry was in its heyday. Management was primarily concerned with simply meeting the high level of demand that existed in most markets. Stockouts and shortages were considered unacceptable, yet adequate logistical systems were unavailable. Manufacturing plants were universally plagued by late deliveries from vendors, poor-quality materials and products, and innumerable disruptions and inefficiencies within the production process itself. Huge stockpiles of raw materials and component parts were the order of the day. High levels of work-in-process inventories were used to reduce the effect of disruptions on the production system in order to more fully utilize workers and equipment. The resulting long manufacturing lead times necessitated large stocks of finished goods to help ensure acceptable levels of customer service. In all areas of the typical plant, massive amounts of inventory were used as protection against the uncertainties of the manufacturing environment. But it eventually became apparent that this mode of operation was unacceptable.

Traditional Approaches

The beginnings of modern manufacturing control were first conceived within this high-inventory environment. Concepts such as statistical order point and economic order quantity became the buzzwords of production and inventory control managers. These decades-old, cost-based inventory replenishment strategies were embraced as the state-of-the-art techniques for inventory control. Even though these techniques improved operations in some areas, it became increasingly evident that these concepts had limited applicability in most manufacturing environments.

The basic elements of an integrated planning and control system were first identified and defined during the sixties and continued to evolve throughout the decade of the seventies. [18, 19, 20] One concept, material requirements planning (MRP), became the focal point of a national crusade to improve manufacturing operations. MRP can be defined as a set of techniques that uses bills of material, inventory data, and the master production schedule to calculate requirements for materials. It makes recommendations to release replenishment orders for material and to reschedule open orders when due dates and need dates for these orders are out of step with each other. The MRP crusade helped shift the emphasis away from the traditional "just-in-case" inventory mentality and toward a manufacturing control system based on actual need dates and quantities. Although originally seen as merely a way to order inventory, MRP has been widely used as a scheduling technique, i.e., a method that attempts to establish and maintain valid due dates on orders.

Unfortunately, MRP never seemed to live up to all of the promises of the MRP crusaders. Experience taught us all too well that MRP had a number of inherent weaknesses and was not the final solution to our problems. In the attempt to overcome the shortcomings of MRP, U.S. industry has spearheaded a continuing evolution of the MRP concept, with each additional evolutionary phase giving rise to ever greater unfulfilled expectations.

Closed-loop MRP developed from attempts to provide updated feedback to the MRP system through production planning, master production scheduling, and capacity requirements planning. Once the planning phase is complete and the plans have been accepted as attainable, the execution functions come into play. These include the shop floor control functions of input-output measurement, detailed scheduling and dispatching, plus anticipated delay reports from both the shop and vendors, purchasing followup and control, etc. The term *closed-loop* implies that not only is each of these elements included in the overall system but also that there is feedback from the execution functions so that the planning can be kept valid at all times.

Integrated Manufacturing Systems

The evolution of thought concerning manufacturing control continued in the mid-seventies with the emerging concept of the integrated manufacturing system. MRP enthusiasts helped to obscure the essential issues by coining the term *MRP II* as a title for the integrated systems approach known as manufacturing resource planning. Unfortunately, this terminology served to keep the emphasis on MRP instead of on the integrated system.

MRP II is a system designed for the effective planning of all resources of a manufacturing company. MRP II systems have three major characteristics.

In addition to being a complete closed-loop system, including the planning and control of priority and capacity, it also integrates production and inventory plans into the financial planning and reporting system. Furthermore, it has a simulation capability to answer "what if" questions.

The Report Card on MRP Systems

According to George Plossl and Raymond Lankford, MRP systems have been oversold to U.S. industry to the extent that "in a classic case of fad psychology, companies avidly embraced MRP as the solution to their problems, oblivious to the other aspects of manufacturing control essential to making MRP work." They further contend that, based on the companies that have implemented MRP II systems, "MRP II is among the most overemphasized and underachieved goals of U.S. industry." [25]

A study conducted by David Whiteside and Jules Arbose, published in 1984, indicates that "Companies using MRP and other computerized production planning systems have preserved high levels of inventories as usual." Furthermore, the authors state that some critics believe that "MRP is a $100 billion mistake, and 90% of MRP users are unhappy." [26]

Today, MRP-based systems are still undergoing their initial adoption by many of our largest companies and are just now filtering down to many of our smaller production plants. It is believed that as many as 5,000—and perhaps more—U.S. companies are using MRP today. Most of these users report annual sales exceeding $20 million. It is unfortunate that only a small minority of these MRP adopters to date have become highly successful users.

However, the news is not all bad. MRP/MRP II are demanding systems that require tremendous amounts of information, extremely accurate record keeping, up-to-date manufacturing data files, and strict discipline on the shop floor. Successful implementation of MRP usually requires the development of new lines of communication and improved auditing procedures within the firm. Those firms that have benefited the most from MRP have typically done so because of a strong commitment to improved data accuracy and plant discipline. In a sense, MRP has helped these firms by forcing them to go back to the basics and do some of the things they should have been doing all along.

Even firms that have successfully implemented MRP systems recognize that there is still significant room for improvement. They realize that something is missing. Evidence of lingering manufacturing inefficiency is exemplified by the continuing occurrence of late orders, excessive expediting, wandering bottlenecks, excess work in process, and invalid schedules. It has become all too clear that neither MRP, closed-loop MRP, nor MRP II have the capability to generate valid schedules in a manufacturing environment.

MRP uses excessively long, predetermined lead times to backward schedule from established due dates without taking into consideration the interaction of products competing for the same resources. Moreover, MRP systems simply computerize existing production practices. Unfortunately, most managers failed to recognize that the old-line manufacturing practices contributed greatly to the inability of firms to reach their competitive potential. The typical firm implementing MRP failed to focus on actually improving the product flow of the plant. This is a fatal flaw in the basic MRP approach that virtually guarantees unnecessarily high inventories and unreliable schedules.

An appropriate question at this point is, "What alternative does management have?" Should the MRP approach be junked in favor of something different? Absolutely not! In essence, MRP systems are simply information systems that are capable of supplying the information required to make internal or external decisions across different functional areas and through time. As such, MRP systems provide excellent information to support manufacturing operations. Many functional areas work more efficiently because of the information supplied by MRP.

The synchronous manufacturing philosophy, which is developed in this book, includes a state-of-the-art manufacturing control system that focuses attention on the key constraints and control points in the plant, while requiring less data than an MRP system. However, much of the information contained in MRP-type databases is very valuable and can be used in the implementation of a more effective manufacturing control system. Thus, MRP systems, where they exist, should be used to support the move toward a synchronous manufacturing philosophy.

IMPROVING THE MANUFACTURING LOGISTICAL SYSTEM

The Role of Manufacturing Management

The fundamental approach to attaining excellence in any manufacturing system is through the improved management of that system. Along this line, there is much that can be learned from the Japanese philosophy of manufacturing management. A number of firms have copied and applied parts of the Japanese management philosophy to their own operations with some degree of success. But it is not reasonable to expect that the Japanese system of manufacturing management can be totally copied by firms that exist within other cultures. Companies that try to emulate the Japanese may be able to improve their operations. But in the final analysis, the approach of trying to copy as much of the Japanese system as possible dooms the firm to long-term inferiority to the Japanese. The ability to successfully compete with the Japanese in today's environment hinges on whether management can develop an equal or superior manufacturing logistical system that contains a mechanism for continual improvement within the totally integrated firm. But before management attempts to improve their manufacturing logistical system, they must first understand the evaluation and control processes that are used to monitor and drive that system.

The Role of Management Accounting

Management accounting can be described as the process of identification, measurement, accumulation, analysis, preparation, interpretation, and communication of financial information used by management to plan, evaluate, and control within an organization. Management accounting is also responsible to ensure appropriate use of, and accountability for, its resources. Accounting thus records all business activity, analyzes the various parts of the organization, and measures the dollar impact of the activities within each part of the organization. In a manner of speaking, management accounting has always been the original systems integrator and will probably continue to fill that role for the foreseeable future.

The judgments and recommendations of the accounting staff have become the basis for most of the important managerial decisions within the typical firm. Accounting measures and standards are also directly responsible for many of the operational policies and guidelines that exist in manufacturing firms today. Given the importance of the role played by accounting and the impact that it has in our manufacturing logistical systems, standard cost accounting practices must be closely examined.

Dr. Eliyahu M. Goldratt has argued most adamantly and eloquently for years that cost accounting is the number one enemy of productivity. [1, 27] However, Dr. Goldratt no longer stands alone in his convictions, as many others have joined the crusade. According to Harvard University Business School professor Robert S. Kaplan, "Efforts to revitalize manufacturing industries cannot succeed if outdated accounting and control systems remain unchanged. Yesterday's accounting undermines production." [28] Interestingly enough, Dr. Kaplan is the Arthur Lowes Dickenson Professor of Accounting at Harvard University.

What has caused this outcry against the traditional accounting practices? The basic argument is that the standard cost procedures and the performance measures supported by these cost systems all too often trigger dysfunctional actions within the organization in general and specifically within the manufacturing system. The reason for the occurrence of these dysfunctional actions is that traditional cost accounting systems try to maximize the efficiency of individual subsystems instead of optimizing the performance of the total system.

An all too familiar illustration of applying ill-conceived accounting efficiencies and measures is the end-of-the-month syndrome. In many manufacturing plants, the shipping records exhibit an interesting pattern. The number of orders actually shipped during the first part of the reporting period (typically one month) occurs at a relatively slow pace. Then, toward the end of the reporting period, the shipping rate increases sharply. This pattern is repeated period after period, creating what some amusingly refer to as a hockey stick effect on shipping charts. The end-of-the-month syndrome is illustrated in Figure 1.1.

The hockey stick phenomenon appears in all sizes of plants and in facilities with widely varying degrees of technology. Although the relative size of the shipping rate difference between the first part of the period and the last will vary from one plant to another, the basic pattern remains. The type of industry, the specific products being produced, and the skill level of the workforce also do not seem to matter. The only parameter that affects the shipping rate pattern is the length of the reporting period. If the reporting period is monthly, then the hockey stick pattern recurs every month. If the reporting period is quarterly, then the pattern cycles every quarter. It is evident that the manner in which data are reported significantly affects our actions in the plant.

All employees, whether they be managers, supervisors, or hourly workers, are influenced by the performance evaluation measures currently used in the plant. Their behavior and performance simply reflect the standards of the existing evaluation system. The relevant questions are, "What actions are encouraged by our cost accounting systems? And are they consistent with the overall goal of the organization?"

FIGURE 1.1   THE END-OF-THE-MONTH SYNDROME

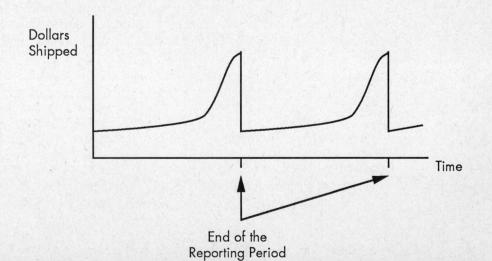

Chapter 1   The Manufacturing Environment Today

The first part of each month, the actions of the workforce are primarily influenced by standard cost accounting performance measurements. These measurements have a local focus. They emphasize the efficiency and utilization of specific machines, workers, work centers, and departments. The measurements stress the standard time to run a part and the per unit cost to produce that part at each operation. As a result, the supervisors act as if they are wearing blinders as they strive to meet the established time and cost standards by running large batches of products without regard to how their actions affect the rest of the plant.

Individual departments working in virtual isolation, each striving to meet their own efficiency standards, can have a devastating effect on downstream operations. Large batch sizes result in excess work-in-process inventories of some items and shortages of others. The excess inventory occurs because large batches result in more product than is needed in the short run. Shortages occur because a lengthy production run of any one item causes a temporary halt to the production of other items that may be required at downstream operations. Thus, some operations may develop huge backlogs of work, while other operations are starved for work and must shut down. Supervisors also create bottlenecks by utilizing only their most efficient machines (in order to keep their efficiencies high), while less-efficient machines sit idle. Bottlenecks appear to move around, even popping up at work centers that normally have a lot of excess capacity. The result is poor work schedule execution and performance for the entire system as shipping falls further and further behind schedule.

As the end of the reporting period approaches, with shipments now clearly behind schedule, the plant manager intervenes. The local performance measures are temporarily displaced by more global performance measures. General financial accounting performance measures for the entire plant such as the income statement take center stage. Thus the immediate concern becomes one of meeting the shipping goals for the reporting period. All energies are now focused on expediting enough of the partially finished orders to meet these shipping goals. Workers are paid premium wages to work overtime. Expeditors initiate actions that result in split and overlapped batches, altered priorities, and changed work schedules. Efficiencies are ignored as products are processed through inefficient machines. Orders are hurriedly completed and shipments temporarily increase. Once the crisis is over and the reporting period ends, the plant again reverts to its local cost accounting measures. The groundwork is laid for the cycle to repeat.

Manufacturing firms in the United States have developed measurement systems that cause our employees, management and workers alike, to make operating decisions that are inconsistent with the overall goal of the organization. We will more closely examine this vitally important concept of performance measures and their effect on productivity in a later chapter. That discussion will detail how the traditional performance measures have negatively affected productivity. More important, an effective system of performance measurement will be presented. But first, the significance of the answer to one basic question must be fully understood. What is the goal of every manufacturing firm?

Understanding the Goal of Manufacturing Firms

With the emergence of a host of new and powerful techniques, it is easy to miss seeing the forest for all the trees. Managers may sometimes wonder exactly what is meant by terms such as manufacturing resource planning, group technology, flexible manufacturing systems, just-in-time systems, total quality control, and computer integrated manufacturing. It is at best a time-consuming chore to understand any one of these concepts, much less to fully comprehend the ramifications of the implementation of one of these systems in a specific manufacturing plant environment. To make the best choice from among the new system technologies is indeed a formidable task. In order to improve the likelihood of making the correct management decisions, one must first recognize the goal of every manufacturing organization.

Is the goal of a manufacturing firm to provide a high-quality product at a competitive price? Is the goal to offer better customer service? Is the goal to capture market share? Is the goal to have the latest technology-based equipment and machinery? Should the goal be to reduce costs? Or should the corporate goal be to survive? While all of the above may be considered to be worthwhile means to achieving the goal, they are not *the* goal of the organization. There is and can be only one goal for a manufacturing firm. The goal is to make money, both now and in the future! [1, p. 40]

REACHING FOR THE GOAL

Every management decision should be based on helping the firm reach its goal of making money. Every investment decision should be based on whether or not that investment will help the firm reach its goal. Therefore, we should try to restrain ourselves from the temptation of buying our way into being productive through massive injections of robots and computer-based equipment. The goal is to make money, not to spend it. Furthermore, in analyzing and choosing between manufacturing logistical systems, we must not choose to simply copy the Japanese. We must develop a manufacturing management system that encourages every individual in the firm to make decisions that contribute to the goal of making money.

The Traditional Approach

The traditional bottom-line financial measurements of making money are net profit, return on investment, and cash flow. Net profit is an absolute measure of whether or not the company is making money. But net profit by itself is inadequate. Suppose a firm realized a net profit of $2 million last year. Is a net profit of $2 million good? That judgment cannot be made without knowing the level of investment required to earn that level of profit. Return on investment is a relative measure of making money, which relates the level of earnings to the level of investment. If the investment was $10 million, then the $2 million represents a return on investment of 20 percent, a nice return. If the investment was $100 million, then the return on investment is 2 percent, which is not very good. The bottom-line measure of cash flow refers to the amount of money available to meet the firm's financial obligations. Cash flow is often overlooked but is nonetheless important. Many executives have learned the hard way that without a sufficient cash flow, the firm cannot survive.

The bottom-line measures of net profit, return on investment, and cash flow are sufficient to indicate whether the firm is currently making money. But they are terribly inadequate when it comes to evaluating operating and investment decisions. Can they help us determine what products to schedule or what batch size to run? Can they indicate whether or not a new order should be accepted? Can they help us decide whether or not to purchase a new piece of equipment? The answer to all of these questions is no. Currently, when these types of decisions must be made, managers use a combination of cost accounting concepts and intuition. [5, pp. 22–23]

As we saw in the end-of-the-month syndrome illustration, the use of cost concepts may very well lead us astray and encourage us to make the wrong decisions. That is why intuition (which comes from experience) traditionally has played a large role in manufacturing decisions. Many managers have learned from experience not to trust what the cost figures tell us. For example, managers have long since learned not to rely on the cost-based economic order quantity (EOQ) formulas to determine what batch sizes to run. We shall see additional examples of how cost concepts can be misleading in later chapters. The fact that intuition and experience constantly cause us to override cost recommendations leads us to search for ways to improve our decision-making process.

The Concept of Synchronous Manufacturing

Managers are painfully aware of the consequences of missed production and shipping goals or excessive operating expenses. Traditional cost accounting performance measures tend to keep management's attention focused on actions that address immediate problems. In order to help solve these short-term problems, managers have frequently relied upon carrying lots of "just-in-case" raw material, work-in-process, or finished goods inventory. The lack of effective logistical systems has been the primary factor in the inordinate lack of control over inventories. Inventory has been used by managers as a "wall" to protect their plants from the uncertainties, complexities, and disruptions of the manufacturing environment. But given the realities of competition in the world today, managers can no longer afford to operate with a "business as usual" attitude.

Manufacturing managers are increasingly discovering an important principle that Henry Ford clearly understood nearly 70 years ago—maintaining high levels of inventory ruins a firm's chance at gaining or maintaining a competitive advantage in the marketplace. It is evident that many Japanese manufacturers have mastered and implemented this principle. During the course of this book, it will be clearly demonstrated how excessive inventory, especially work-in-process inventory, adversely affects a firm's manufacturing lead times, material and product flows, and competitive edge.

Managers must learn how to protect their plants from the everyday disruptions of the manufacturing environment without the use of massive levels of inventories. Moreover, they must learn to effectively manage the entire plant as a single unified and synchronized system.

Acknowledgment

The philosophy of synchronous manufacturing must be largely attributed to the many significant contributions of Dr. Eliyahu Goldratt. It is often difficult to accurately reference ideas and concepts that are still in an evolutionary stage, many of which are unpublished at this current date. Therefore, we gratefully acknowledge Dr. Goldratt's many contributions to this new philosophy and to this book. For those readers who are not already familiar with Dr. Goldratt's work, we highly recommend *The Goal* and *The Race* as additional reading. [1, 5]

ORGANIZATION OF TEXT

The purpose of this book is to present the new and important philosophy of synchronous manufacturing. It will expose some of the more critical weaknesses of traditional manufacturing management practices; develop the basic principles required to manage manufacturing operations; explore the significant role that constraints play in a manufacturing firm; examine the principles that provide the foundation for assembly lines and just-in-time logistical systems; develop a universally applicable logistical system consistent with the synchronous manufacturing philosophy; present a new structure for classifying and analyzing manufacturing operations; and illustrate the basic application of synchronous manufacturing with real-life case studies.

This book is the first of two volumes on synchronous manufacturing found in the APICS–South-Western series in production and operations management. The second volume will concentrate on synchronous manufacturing implementation strategies and in-depth case studies. The second volume assumes mastery of the concepts presented in the remainder of this book.

The traditional cost-based procedures of evaluating decisions in manufacturing environments are often ineffective. In Chapter 2, this fact is demonstrated with a numerical example, and a new perspective for evaluating manufacturing decisions is established.

Many of the traditional manufacturing management practices are based on principles that need to be reconsidered. Chapter 3 examines the basic product and resource interactions that characterize manufacturing environments. This process leads to the development of a set of new principles. These new principles lead to a better understanding of how to manage our manufacturing plants.

Constraints determine the limits of performance for all manufacturing organizations. Chapter 4 focuses on this critically important issue of constraints and identifies and analyzes the various types of constraints that exist in our manufacturing plants.

Chapter 5 considers the evolution of the synchronous manufacturing philosophy. The basic concept of a synchronized product flow is explained and additional principles of synchronous manufacturing are developed. Also in this chapter, two very significant logistical systems—the assembly line and the just-in-time philosophy—are analyzed. Their strengths and weaknesses are identified, and their underlying principles are compared with and contrasted to the basic principles of synchronous manufacturing.

The synchronous manufacturing logistical system, identified as drum-buffer-rope, is developed in Chapter 6. This logistical system satisfies all the requirements of the synchronous manufacturing philosophy and has universal application to all types of manufacturing environments. The drum-buffer-rope logistical system carefully controls the flow of materials into and through the plant in an attempt to produce finished goods in accordance with market demand with a minimum of inventory and operating expense. A key consideration in this logistical approach is the identification of all the processing, resource, and market constraints within the entire system. These constraints drive the planning, scheduling, and control of all the plant's resources. The resulting synchronous manufacturing system is one in which there is a smooth and continuous flow of materials moving quickly through the plant with a minimum of disruptions.

Chapter 7 presents a discussion of techniques that can be used to improve the performance of the firm. The discussion primarily involves two different approaches to improving performance. One approach, primarily for production-constrained firms, focuses on methods designed to improve the throughput of the firm. The other approach centers on how to utilize the drum-buffer-rope logistical system to identify those areas of the firm where improvements will have the greatest bottom-line impact.

Chapter 8 recognizes and addresses the complexities that exist in all manufacturing plants, and develops the concept of product flow diagrams so that complex product flows can be adequately analyzed. All manufacturing plants can be classified as either V-plants, A-plants, T-plants, or some combination of these three categories. It is logically developed that plants which have similar structural characteristics also have similar problems. Thus, the V-A-T framework provides the foundation for analyzing manufacturing operations based on their structural classification. Individual case studies are presented for a V-plant, an A-plant, and a T-plant in order to demonstrate the applicability of synchronous manufacturing concepts to real-life manufacturing environments.

SUMMARY

The United States is losing ground to foreign countries in the manufacturing sector. This is vividly demonstrated by numerous economic indicators such as the balance of trade payments and the rate of productivity growth. For some time now, many have tried to blame our industrial decline on labor unions, government regulations, low-cost foreign labor, and other various forces beyond the control of corporate America. But it is now becoming apparent that the main culprit for our loss of competitive position in the marketplace is management.

The development of systems such as MRP II has been a mixed blessing. On one hand, the discipline required to run such systems has forced some firms to improve their database and management planning and control systems. This alone has made a significant difference in the operation of many firms. However, most firms that have implemented MRP systems have fallen far short of the competitive level required to compete successfully in the international marketplace. This failure is largely due to misguided management practices, not to inherent flaws in the MRP system.

To regain the competitive edge, managers of manufacturing organizations must reexamine traditional managerial philosophies and systems, particularly in the area of planning and control. In fact, the entire focus of manufacturing activities should be changed from the local view encouraged by the traditional management accounting practices to one that is in tune with the global and competitive needs of the business and its goal to make money now and in the future. Synchronous manufacturing is the systematic approach to accomplishing this objective.

QUESTIONS

1. Why are "walls" often ineffective?
2. In what sense is inventory a "wall"?
3. Explain how "walls of inventory" are dysfunctional.
4. What are some of the economic indicators that suggest that the United States is losing its competitive edge in manufacturing? What are some additional indicators not included in this chapter that further substantiate the proposition that U.S. manufacturing is in trouble?
5. Explain the meaning of the "hollow corporation." What are its implications to the long-term future of the national economy?
6. Is a pure service sector economy viable? Explain why or why not.
7. Identify some of the traditional arguments of why U.S. manufacturers cannot compete with foreign competitors. Why are these arguments invalid?
8. How was MRP designed to shift a plant away from the "just-in-case" mode of production?
9. In what sense has MRP failed?
10. Why do traditional management accounting practices lead to poor decisions? Would the "Quality Is Job 1" program at Ford Motor Company and the "Saturn" program at General Motors fall within the standard justification procedures? Explain.
11. Describe and explain the end-of-the-month syndrome.
12. What is the goal of a manufacturing firm, a bank, and the U.S. Postal Service?
13. Why do organizations need a clearly defined quantitative goal?
14. Identify and explain the financial measures that are traditionally used to evaluate a firm's performance.

2

Manufacturing Decisions—A New Perspective

INTRODUCTION

The end-of-the-month syndrome presented in Chapter 1 provided a graphic illustration of how traditional cost procedures and performance measures can distort our perception of what constitutes appropriate manufacturing decisions. Unfortunately, poor decisions caused by the misuse of traditional cost procedures and guidelines are not limited to just one aspect of the manufacturing function. The ultimate impact of specific actions by functional areas, such as purchasing, engineering, and marketing, on the financial well-being of the firm are also distorted by the use of traditional cost methods. In fact, in most firms, the bottom-line financial effect of most functional decisions is very difficult to determine. As a result, normal adherence to traditional cost procedures makes it difficult to establish effective management policies and operating procedures. But with a new perspective, it is possible to determine accurately the financial impact of specific management decisions in the manufacturing environment.

In this chapter, a systematic approach is developed that will provide the foundation for calculating the impact of planned activities on the entire operation. This approach to analyzing the impact of management decisions will enhance the ability of managers to make more effective operating decisions.

But it is extremely important to understand first why our traditional cost procedures and managerial practices fail to lead managers to the proper decisions. This understanding is essential since the use of cost standards and performance measures is at the heart of almost all aspects of manufacturing decision making. These cost guidelines are used in purchasing and materials management, in the scheduling of workers and machines on the factory floor, in pricing and marketing strategies, and in establishing how people in the organization are motivated, measured, and rewarded. Without a firm grasp of the inadequacies of our current procedures, managers can expect to stumble only occasionally upon the correct decisions. Once the reasons for the failure of the traditional cost procedures are exposed, basic principles will be developed that will serve as a guide to formulating a more effective global approach to managing manufacturing operations. These principles provide the foundation for synchronous manufacturing.

LOCAL OPTIMA VERSUS GLOBAL OPTIMUM

The goal of managing a manufacturing operation is to maximize the financial performance of the entire system. This goal is clearly understood by most business managers. The problem is the lack of appropriate guidelines that can help managers achieve the goal. Consequently, the history of American industry is littered with the carcasses of unsuccessful companies. Many of these companies failed because of cost procedures that left their managers helpless to properly evaluate the ultimate systemwide consequences of a variety of critical decisions.

Every manufacturing business is composed of various subsystems. First, the product has to be designed and engineered. Then, the required materials and resources must be purchased. Finally, the product has to be manufactured, marketed, and the revenue collected from the buyer. It is not desirable for the firm to try to optimize the operation of each of these individual functional areas because their objectives are often incompatible. For example, in a typical manufacturing firm, the sales staff always prefers their product line to be as full as possible with a large number of options. But accounting emphasizes the cheapest way to produce the product. And production would always opt for the easiest way to manufacture the product. If the firm is to be successful, all of the various subsystems must be in harmony with the common goal of the entire system. The local optima that might be achievable in each subsystem must be subservient to the global optimum for the total system.

In manufacturing organizations, it is not an easy task to estimate the impact of policies or decisions within one subsystem on the total system. As the organization grows and each subsystem increases in complexity, it becomes necessary to subdivide the subsystems because of span of control limitations. As a result, each subsystem becomes a less significant part of the total system, and it becomes increasingly difficult to determine the impact of local departmental decisions on the entire system. To illustrate the point, consider the mature manufacturing system that is often subdivided into a number of subsystems such as purchasing, receiving, and shipping, as well as various production departments. In this type of environment, it is not a trivial matter to estimate the impact of a batch-size decision for a single product at a particular machine on departmental performance. However, it is an even more formidable task to evaluate the impact of batch-size decisions on the efficient operation of the manufacturing subsystem. Finally, without precise and appropriate guidelines, it is virtually impossible to evaluate the impact of batch-size decisions on the organization as a whole.

In most manufacturing plants, numerous operational decisions are made every day that greatly impact the firm's productivity and profitability. But the traditional methods of analysis used by most organizations today do not provide managers with the tools necessary to develop consistently good solutions to their problems. This chapter clearly demonstrates that manufacturing organizations must carefully reconsider their decision-making processes. This includes a critical examination of key assumptions and procedures that may lead to faulty analysis and costly managerial decisions.

THE STANDARD COST SYSTEM

What guidelines can be used in order to make sound decisions within complex manufacturing organizations? The traditional approach of most organizations has been to rely on what is commonly referred to as the standard cost system.

The standard cost system has been the driving force behind most management control and decision-making processes in U.S. manufacturing firms for four decades. Included in this approach are detailed procedures for calculating the cost impact of any proposed action. For many years, the standard cost system has survived as the normal mode of operation. Despite its many limitations, a better alternative was simply not available.

But the increasingly competitive manufacturing environment has changed everything. Now there is a growing awareness among those in the accounting profession that the standard cost system is in need of an overhaul. [6] This system has a number of fundamental flaws that hinder, rather than help, management in their struggle to become more competitive. Two major problems of the standard cost approach examined in this chapter are:

1. The cost system is based on assumptions that are invalid.
2. The cost system is local in scope and strives to reduce the cost of each process and product in isolation. In attempting to achieve these local optima, the cost approach actually encourages a system that is far removed from the global optimum.

Assumptions of the Standard
Cost System

There is a widely held belief that the role of manufacturing management is to minimize the cost at each individual operation. In fact, the general approach to cost-based systems has been to assign to each operation a cost or financial measure of its impact on the total system. Then the standard system is constructed in such a way that:

1. Total cost of the system = Sum of cost at each operation

In this approach, each manufacturing subsystem affects the total business by contributing to the cost of manufacturing the products.

To better understand the inherent shortcomings of the traditional cost approach, one must first realize how costs are assigned to each operation. In the general cost approach, all of the overhead expenses of the system must be allocated to the mix of products produced by the system. Most allocation procedures estimate the total dollars of overhead cost, divide by estimated direct labor cost, and allocate burden to the product's cost as a percentage of direct labor dollars. [29]

There are numerous other procedures that can be utilized to allocate burden. However, the problems discussed here are not the result of using one basis for allocation instead of another. The basic problem is in the allocation of burden itself and the assumptions that naturally follow any such allocation.

The standard cost procedure most widely used (which is based on allocating burden according to the direct labor cost) prescribes that:

2. The total cost at each operation is proportional to the cost of direct labor for that operation.

And also that:

3. The total cost for the system (excluding material cost) is proportional to the sum of the direct labor costs.

The point to be emphasized here is that the traditional cost approach is set up to satisfy the basic accounting requirements expressed by the first equation. Therefore, an overhead proportionality factor that reflects the historical overhead and burden rate is determined. Only then can a cost for each product on each process be assigned. However, this standard cost scheme is useless as a decision-making tool unless we make an additional assumption:

4. The standard cost procedure, which utilizes the calculated overhead/labor ratio, can be reversed to estimate the impact of any action on the total cost of the system.

Implications of the Standard Cost System

The consequences of applying the assumption in item 4 are extremely serious. The entire structure of managing manufacturing operations by traditional cost systems hinges on this assumption. It implies that if the direct labor cost of a given operation can be reduced, then the total cost allocated to that operation will be reduced by a proportionally larger amount. It also implies that the total cost for the whole system will be reduced by a similar amount. Acceptance of this assumption further implies that the standard cost calculation can be used to accurately determine the financial impact of any local action on the entire system. Thus, in one giant, oversimplified step, our attention is focused on the management of cost at each individual operation, within each separate department, or in each functional area.

Responsibilities can now be clearly delineated. Each manager is responsible for controlling the cost of the operations that fall under his control. Engineers can now focus attention on specific processes or products and use the standard cost calculations to compute benefits. Without such a systematic procedure, engineering decision making would be impossible. Since the cost to manufacture each product can be calculated, marketing is apparently able to perform its task effectively. Marketing can focus on selling products with high margins and stay away from products for which the selling price is less than the calculated manufacturing cost. The entire puzzle seems to fall neatly into place.

Notice, however, that the basic assumptions inherent in the standard cost procedure ignore the effect of interactions between the various elements of a subsystem and the effect of interactions between the various subsystems. This is particularly true of the assumption in item 4. The standard procedure assumes that if the cost of a single product or process is decreased (or increased), then the total cost for the system will be decreased (or increased) by a proportional amount. Implicit in this reasoning is the following conclusion:

5. *In manufacturing operations, the effect of optimizing local decisions (as measured by their impact on the cost of the operation) is to optimize the total system.*

To put it more succinctly, we operate our businesses with a firm belief that the following motto is true:

6. *The key to reaching the global optimum is achieving local optima.*

Limitations of the Standard Cost System

Of course, on an intellectual level, it is understood that the real world is not this simple. The manufacturing environment is an extremely complex system of resource and product interactions. It is also recognized that the inherent assumptions of the cost system are probably not valid. Nevertheless, the guiding principles used by managers in many of our manufacturing organizations are anchored in the localized viewpoint based on traditional cost systems. Many of the problems that plague American industry today are due largely to the effects of this approach. These problems are a direct result of the view that the goal of manufacturing management is to control and reduce the standard cost of each individual operation.

An Illustration   Suppose the plant manager of a manufacturing firm is considering a proposal to purchase a new and faster stamping machine. The basic information available to analyze the purchase decision is presented here:

- The old machine is able to process material at a rate of 100 units per hour (one unit every 36 seconds).
- The new machine is three times as fast, processing material at a rate of 300 units per hour (12 seconds per unit). This saves 24 seconds, or .006667 hours, per unit processed.
- Approximately 150,000 units per year are processed at the work station where the stamping process is performed.
- The cost of direct labor is $15 per hour.
- The overhead factor for this process is 2.80 times direct labor.
- The net cost of the new machine is $45,000. (This includes the salvage value of the old machine.)

The expected savings from the purchase of the new machine would typically be calculated in the following way:

$$
\begin{aligned}
\textit{Direct annual} \quad &= \textit{(reduction in processing time)} \\
\textit{labor cost savings} \quad & \\
&\times \textit{(number of pieces produced per year)} \\
&\times \textit{(cost of direct labor)}
\end{aligned}
$$

The direct labor cost savings are then converted into total cost savings through the use of the appropriate overhead factor:

$$
\begin{aligned}
\textit{Overhead cost savings} &= \textit{(direct labor cost savings)} \\
&\times \textit{(overhead factor)}
\end{aligned}
$$

And the total annual cost savings are calculated as follows:

$$
\begin{aligned}
\textit{Total annual cost savings} &= \textit{Total annual labor cost savings} \\
&+ \textit{Total annual overhead cost savings}
\end{aligned}
$$

While the exact procedure for the above calculations may differ between firms, the fundamental approach is the same.

Continuing with the illustration of the stamping machine:

$$
\begin{aligned}
\textit{Direct labor cost savings} \quad &= (.006667 \ \textit{hours per unit}) \\
&\times (150{,}000 \ \textit{units per year}) \\
&\times (\$15 \ \textit{per hour}) \\
&= \$15{,}000 \ \textit{per year} \\[4pt]
\textit{Total annual overhead} \quad &= (\$15{,}000) \times (2.80) \\
\textit{cost savings} \quad & \\
&= \$42{,}000 \ \textit{per year} \\[4pt]
\textit{Total annual cost savings} &= \$15{,}000 + \$42{,}000 \\
&= \$57{,}000
\end{aligned}
$$

Since the net cost of the new machine is $45,000, the direct labor cost savings of $15,000 translates to a payback period of 3 years. In many firms, the investment could be approved on that basis alone. Considering both indirect and direct costs, the total annual cost savings for the department are $57,000 per year. From a local point of view, this implies a payback period of less than 1 year. Thus, from a standard cost perspective, the purchase of the stamping machine would likely be approved.

Realities of the Manufacturing Environment   Now let's consider how the realities of the manufacturing environment might impact the above illustration. It is appropriate to consider three important issues before a final judgment can be made on the correctness of the cost-based purchase decision. These issues are the capacity requirements of the stamping operation, the actual reduction in direct labor, and the allocation of indirect costs.

Capacity Requirements   One major issue in the stamping machine illustration is whether or not the new equipment is actually needed. To address this aspect of the problem, several key questions need to be answered. For example, is the old machine fully utilized? Is there a need to increase the production capacity in the stamping process? Is the old machine inadequate to meet the demands of the process? Upon further investigation, suppose it is discovered that the old machine is utilized 75 percent of the time. Does this additional information support the purchase decision? It is difficult to jump to that conclusion. The real determination that must be made is whether or not there is a need for additional processing capacity. And this question can only be answered from the point of view of the whole system.

This discussion raises the following dilemma. If there is not a need for additional processing capacity, then it is difficult to justify the investment. On the other hand, if there is a need to increase the output of the stamping process, then even with a faster machine, it is unlikely that the total direct labor cost of the operation will be reduced. Even if the direct labor cost can be reduced, it is also unlikely that it will be reduced proportionally to the assumed reduction by the standard cost procedures. (The inability to reduce the direct labor is compounded by the fact that installing a faster machine usually results in significantly lower gains than expected in processing efficiency.) [30] In either case, the cost justification calculations that have been made are not valid!

Reduction of Direct Labor   Another key issue is whether or not labor can actually be reduced even if the utilization of the new machine is less than the old machine. Typically, if the labor content of the operation is to be reduced, then one or more workers must be removed from the process. In many cases, the difference in labor requirements from the purchase of a new machine is not sufficient to remove a whole worker from the process. In this case, there is no reduction in the labor content. Moreover, some firms have a "no lay-off" policy. This means that even if a worker can be removed from one operation, that worker will simply be transferred elsewhere within the system. In this case, the labor content has been reduced in the stamping department. But the total amount of labor in the system has not been reduced. From the global perspective, has there been any savings in the cost of labor?

Allocation of Indirect Costs   The final issue raised by the stamping machine illustration is the proper allocation of indirect costs. Consider the cost-of-product calculations for the stamping operation, and ultimately, the products that are processed by the stamping operation. The standard cost analysis would imply that the cost of products that are processed through the stamping operation would be reduced by a total annual cost of $57,000. When divided by the number of units processed annually (150,000), it is concluded that the product cost for these goods is reduced by $.38 per unit. But is such a reduction in product cost actually realized? The answer is a resounding no!

Part of the explanation for the nonrealization of estimated reductions in product costs is based on the arguments expressed in the previous paragraphs. But the primary reason is the way in which indirect costs are calculated and allocated. The procedures used to allocate indirect costs to products are highly questionable. But the important point is that the indirect costs don't go away. Instead, the indirect costs are simply shifted from one product group to another. As a result, the product costs of all products that are not processed through the stamping operation will go up. Even though these products are no less efficiently produced than before, their calculated costs are increased. This is the inevitable result of using cost procedures that have only a local instead of a global perspective.

Efficiency and Utilization

The traditional concepts of efficiency and utilization evolved as part of a manufacturing management framework consistent with the standard cost system. The *APICS Dictionary* [4] defines these two terms as follows:

Efficiency—Standard hours earned divided by actual hours worked. Efficiency is a measure of how closely predetermined standards are achieved. Efficiency for a given period of time can be calculated for a machine, an employee, a group of machines, a department, etc.

Utilization—A measure of how intensively a resource is being used. It is the ratio of the direct time charged for production activities (setup and/or run) to the clock time scheduled for those production activities for a given period of time.

utilization = direct time charged / clock time scheduled

For example, to calculate labor utilization, the direct labor hours charged is divided by the total clock hours scheduled for a given period of time. Similarly, to calculate machine utilization, the total time charged to creating output (setup and/or run time) is divided by the total clock hours scheduled to be available for a given period of time.

The concepts of efficiency and utilization are typically applied as performance measures for various workers, groups of workers, machines, or departments within a plant. This widespread approach to measuring performance clearly assumes that higher efficiency or utilization measurements are better than lower ones. Moreover, this approach assumes that both plant and firm performance are maximized when all entities within the organization are operating at maximum efficiency and utilization. It should be evident that the use of efficiency and utilization measures is clearly consistent with the standard cost system.

The problem with efficiency and utilization measures as traditionally used is that they focus on local performance without regard to how the specific resource interacts with the rest of the manufacturing system. As a result, the use of efficiency and utilization measures often encourages actions that adversely affect the overall performance of the plant. The fact that all resources in a manufacturing plant should not, and typically cannot, be activated to the maximum possible efficiency and utilization will be developed in detail beginning in Chapter 3.

EVALUATING MANUFACTURING ACTIONS—A NEW APPROACH

The standard cost system is the fundamental decision-making technique currently utilized by manufacturing firms. But, as we have shown, the standard cost system has several severe limitations. The primary problem is the lack of a global, or systemwide, perspective on the impact of specific actions. Furthermore, it is incorrectly assumed that calculated local cost savings represent savings for the whole plant.

A Set of Operational Measures

It is time to consider a new approach, one that can properly evaluate the effect of manufacturing actions on the productivity and profitability of the entire firm. The foundation for this new approach is the development of a set of intuitive measures that can be used to evaluate correctly the impact of manufacturing actions. [1, pp. 59–60] These measures will also play a key role in the development of procedures that result in good operating decisions.

In order to be universally applicable, the chosen measures must be common to all manufacturing processes. The measures should also accurately describe the key activities that govern a plant's performance. There are three such basic activities: [2, p. 30]

- The sale of finished products.
- The purchase of raw materials and component parts.
- The transformation of material into finished products.

Based on these three activities, the operational measures of throughput, inventory, and operating expense are now defined. Each of the three measures is defined in a way that differs significantly from traditional definitions. These deviations from the commonly accepted definitions are intentional, and the wording is very precise in order to convey the intended message.

Throughput—The quantity of money generated by the firm through sales over a specified period of time.

The standard definition of throughput is total output. Our definition considers only sales, not units produced. The reason? Finished goods do not generate revenue for the firm until they are sold. Despite the traditional accounting practice of considering inventory as an asset, finished goods are of no value unless they are sold. Throughput is equal to sales revenue minus the material cost of goods sold.

Inventory—The quantity of money invested in materials that the firm intends to sell.

This definition of inventory is simple, yet powerful. The traditional concept of inventory includes the value-added elements of labor and overhead as the material progresses through the transformation process. But the concept of inventory developed here rejects these complicating cost concepts that only distort our perspective and lead to counterproductive strategic and operating decisions. Consider, for example, what happens when material for which there is no demand is processed. This material will eventually be stored as work-in-process inventory. In most plants, material loses flexibility as it is processed. And any loss of flexibility means the material has become less valuable to the firm. Instead of a value-added process, the material may actually experience what we refer to as a value-subtracted process. Thus, in the synchronous manufacturing philosophy, in contradiction to the value-added concept, the value of inventory is always equal to the original value of the material.

> ***Operating Expense***—*The quantity of money spent by the firm to convert inventory into throughput over a specified period of time.*

Operating expense includes all of the money spent by the system, with the single exception of inventory purchases. All value-adding activities (including both direct and indirect labor) are operating expenses. Inventory carrying costs are also considered to be operating expenses. This approach eliminates the confusion about what constitutes an investment and what constitutes an expense.

Notice that money is the common thread in each of these definitions. The relationship of the three is easy to visualize when considered from that perspective. Throughput is the money coming into the system. Inventory is the money currently invested inside the system. And operational expense is the money paid out of the system. Figure 2.1 illustrates the monetary flows represented by throughput, inventory, and operating expense in the firm.

The Effect of Operational Measures on the Bottom Line

There are three factors that account for the power of these three operational measures. First, all three measures are intrinsic to every manufacturing process. Second, the measures are straightforward, easy to understand, and easy to

FIGURE 2.1   HOW MONETARY FLOWS IN THE FIRM ARE REPRESENTED BY THROUGHPUT, INVENTORY, AND OPERATING EXPENSE

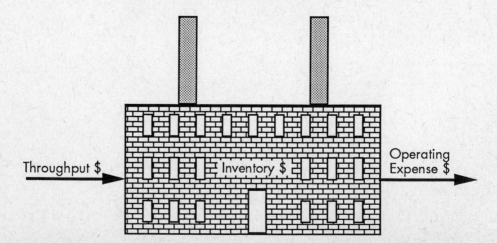

apply. Third, once the impact of any action on these measures is calculated, the impact on the bottom-line measures can also be determined.

All firms ultimately measure financial success according to net profit, return on investment, and cash flow. Thus, the relationships between the operational measures of throughput, inventory, and operating expense and the bottom-line measures of net profit, return on investment, and cash flow are now examined in detail.

When decisions are made that result in the improvement of all three of the bottom-line measures, then the firm is clearly moving in the right direction. Most managers intuitively recognize the relationship that exists between the financial measures and the traditional concepts of throughput and operating expense. The bottom-line impact of inventory in a manufacturing environment is a more complex issue, and one that will receive a considerable amount of scrutiny. But the new operational measures, as defined above, will help clarify and focus the impact of manufacturing actions on the bottom-line financial measures. [2, pp. 32–37]

Throughput and the Bottom Line   It is widely recognized that any action that increases throughput (sales, not production), without an adverse effect on inventory or operating expense, results in an increase in all three bottom-line measures. Under the old definition of throughput (all units produced, whether sold or not), that relationship does not always hold. In fact, an increase in output unaccompanied by an increase in sales will necessarily result in a decrease in all three bottom-line measures.

Inventory and the Bottom Line   How does inventory impact the bottom-line measures? Traditional accounting procedures treat inventory as an asset on the balance sheet. But is it really an asset? The Japanese do not view it as such. In fact, they consider inventory to be "the root of all evil in a manufacturing firm." These two viewpoints are quite contradictory. Later in this chapter, and also in Chapter 5, the significant role that inventory plays in a manufacturing firm will be more closely examined. But for the moment, it is sufficient to understand that actions focusing on inventory reductions in a manufacturing firm can be very beneficial.

At this point, only the direct effects of inventory on the bottom line are examined. Consider the impact of a reduction in inventory without a corresponding increase in operating expense or decrease in throughput. This results in a direct decrease in the level of material investment in the system. Accordingly, since the level of investment is less, the return on investment must increase. Since the amount of cash invested in inventoried material decreases, there also is an automatic increase in cash flow. However, the reduction in inventory does not *directly* affect net profit since throughput is unchanged. Remember that the cost of carrying inventory would be considered an operating expense. Thus, inventory reductions only *indirectly* affect net profit because of the reduced operating expense.

Operating Expense and the Bottom Line   Managers understand that actions which decrease operating expense, without adversely affecting throughput and inventory, have a positive impact on the bottom-line financial measures. However, under the old cost system, trying to sort out the changes in operating expenses is a rather inexact science. But using the new definition, operating expense includes all money spent for anything except the purchase of materials and other assets which can be converted into throughput. It is therefore quite easy to identify when a change in operating expense occurs within a given time period.

An Illustration   The following hypothetical situation demonstrates exactly how the new operational measures relate to the bottom-line measures of net profit, return on investment, and cash flow. Suppose a manufacturing firm has the following year-end financial structure:

Net Sales . . . . $100 million
Cost Of Goods Sold . . . . $90 million
Total Assets (Including Inventory) . . . . $70 million
Inventory (Book Value) . . . . $15 million
Inventory (Material Value) . . . . $10 million
Material Content Of Finished Goods . . . . 40% of net sales, or 44.44%
of cost of goods
Annual Cost to Carry Inventory . . . . 20% per year

Based on the given financial structure, the following values are known:

Throughput	= Dollars generated through sales − Material cost of goods
	= $100 million − (.40 × $100 million)
	= $60 million
Operating Expense	= Total cost of goods − Material cost of goods
	= $90 million − (.40 × $100 million)
	= $50 million
Inventory	= Dollars invested in items intended for sale
	= $10 million
Net Profit	= Throughput − Operating expense
	= $60 million − $50 million
	= $10 million
Return on Assets	= (Net profit)/(Total assets)
	= ($10 million)/($70 million)
	= 14.29%
Cash Flow	= Available cash
	= $10 million

It will now be demonstrated that the impact of changes in the three operational measures on the financial bottom line can be easily determined. In the following illustrations, throughput will increase by 5%, operating expense will decrease by 5%, and inventory will decrease by 10%. First, the bottom-line impact of each of these three changes will be evaluated in the absence of any change in the other two measures. Then, the impact of all three changes occurring simultaneously will be examined. As the illustration develops, it will be interesting to note the relative impact of these changes on the bottom line.*

Case 1: Assume that throughput increases by 5%, while operating expense and inventory remain unchanged.

Since operating expense is unchanged, throughput and net profit both increase by 5% of $60 = $3. Net profit increases to $13.

Return on assets is now calculated for the new net profit figure of $13 using the same asset base of $70. Thus, return on assets is ($13)/($70) = 18.57%

Cash flow is simply the measure of available cash, which has also increased by $3, to a total of $13.

Case 2: Assume that operating expense is decreased by 5%, while throughput and inventory remain unchanged.

A 5% reduction in operating expense translates to a dollar savings of 5% × $50 = $2.5. The new operating expense is $50 − $2.5 = $47.5. Thus, net profit is $60 − $47.5 = $12.5.

Return on assets is now calculated as the new net profit of $12.5 divided by the unchanged asset base of $70. The new return on assets figure is 17.86%.

Cash flow is again increased by the same amount as the increase in net profit, by $2.5, to a total of $12.5.

Case 3: Assume that inventory is decreased by 10%, while throughput and operating expense remain unchanged.

If inventory is cut by 10%, then the asset base is reduced by 10% of the inventory book value. This translates to a reduction of .10 × $15 = $1.5. Operating expense will be indirectly affected since the cost of carrying inventory will be reduced. Since carrying cost is 20%, operating expense is reduced by .20 × $1.5 = $.3.

Net profit increases by the amount of the reduction in operating expense, which is $.3. The new net profit is $10.3.

Return on assets is now calculated as a function of the new net profit of 10.3, and the new asset base of $70 − $1.5 = $68.5. Thus, return on assets is ($10.3) / ($68.5) = 15.04%.

Cash flow is bolstered by the increase in net profit, as well as by the reduction in inventory. The increase in net profit has already been calculated as $.3. This additional cash is a result of the savings in carrying cost and will show up every year. The actual cash freed up by the reduction in inventory is based on the material cost of the inventory. A 10% reduction from the inventory material value of $10 results in a cash savings of

---

*All dollar amounts are in millions.

$1. This $1 in cash savings from reduced purchases of material is a one-time event, occurring only in the first year. But in the first year, the cash flow will increase by $1.3 to a total of $11.3.

Case 4: Assume that throughput is increased by 5%, operating expense is decreased by 5%, and inventory is decreased by 10%.

If all of these changes take place, then net profit will increase by $3 + $2.5 + $.3 = $5.8, for a new total net profit of $15.8. Return on assets will compare the net profit of $15.8 to the new asset base of $68.5, yielding a return on asset figure of 23.07%. Cash flow for the first year will increase by a total of $3 + $2.5 + $1.3 = $6.8.

Implications of the Illustration   It should be made absolutely clear that the operational measures of throughput (T), inventory (I), and operating expense (OE) are not replacements for the bottom-line financial measures. But they are an effective bridge by which to measure the impact of manufacturing actions on the bottom line. In the synchronous manufacturing approach, the appropriate question is not what the cost savings are for a specific action. Instead, the appropriate question is, "What is the impact of the action on the operational measures of T, I, and OE?" It has been shown that the impact of specific actions on T, I, and OE will ultimately determine the financial impact on the entire plant.

Productivity must be measured from the perspective of the entire operation. The standard cost-based analysis falls far short of meeting this requirement. But using the newly presented operational measures, the productive potential of any manufacturing action can be evaluated.

For ease of comparison, the impact of the changes in both absolute and percentage terms for all four cases are summarized in Tables 2.1 and 2.2, respectively. These tables clearly indicate that a relatively small improvement in either throughput or operating expense can be leveraged into a huge increase in the financial well-being of the plant. The data also show that the results of changes in T and OE are several times greater than a reduction in I of twice the size. Since the level of inventory does not directly affect net profit, are inventory reductions less significant than increases in throughput or cuts in operating expense? Unfortunately, that seems to have been the traditional viewpoint. Managers historically have given primary emphasis to productivity programs that focus on total operating expense or throughput, while downplaying the importance of inventory reductions. However, in the next section, as well as in future chapters, it will be demonstrated that inventory considerations are essential to the efficient and synchronous operation of a manufacturing firm.

GAINING THE COMPETITIVE EDGE

The operational measures of T, I, and OE represent a superior alternative to the standard cost system when evaluating the financial impact of

TABLE 2.1   THE ABSOLUTE IMPACT OF SELECTED CHANGES IN THE OPERATIONAL MEASURES ON THE BOTTOM-LINE MEASURES

ABSOLUTE IMPACT ON THE BOTTOM-LINE FINANCIAL MEASURES (DOLLAR AMOUNTS ARE IN MILLIONS.)			
Change in Operational Measure	Net Profit	Return on Assets	Cash Flow
With No Changes - Original Status	$10.0	14.29 %	$10.0
5% Increase in T	$13.0	18.57 %	$13.0
5% Decrease in OE	$12.5	17.86 %	$12.5
10% Decrease in I	$10.3	15.04 %	$11.3
All Above Changes	$15.8	23.07 %	$16.8

TABLE 2.2   THE PERCENTAGE IMPACT OF SELECTED CHANGES IN THE OPERATIONAL MEASURES ON THE BOTTOM-LINE MEASURES

	PERCENTAGE INCREASE OF THE BOTTOM-LINE FINANCIAL MEASURES		
Change in Operational Measure	Net Profit	Return on Assets	Cash Flow
5% Increase in T	30.00%	29.95%	30.00%
5% Decrease in OE	25.00%	24.98%	25.00%
10% Decrease in I	3.00%	5.25%	13.00%
All Above Changes	58.00%	61.44%	68.00%

manufacturing actions. These operational measures also provide a conceptually sound foundation on which to develop a logical and operationally effective philosophy of synchronous manufacturing. But for now the discussion turns to a related and very important consideration—an analysis of the factors that determine a firm's competitive edge. It will also be demonstrated how a synchronous manufacturing system, as outlined in Chapter 1, is a critical prerequisite to a firm achieving its competitive potential.

The Elements of Competitive Edge

Manufacturing firms continually struggle to obtain or maintain an advantage over their competitors. The basis of competition varies from one industry, company, or product to the next. But firms usually achieve the competitive edge in one of three ways: (1) by having better quality products, (2) by offering superior customer service, and (3) by being the low-cost producer.

Product quality can be measured in several different ways. However, in the context of the manufacturing environment, the most important consideration is product conformance to design standards. This is one area where the Japanese have excelled. They have managed to penetrate markets worldwide because of the excellent reputation of their products in conforming to specified standards.

Customer service is often the deciding factor in selecting a supplier from which to purchase. The best quality product is of little value to a potential customer if the product is not available when needed. Customer service can be measured in two different ways. One aspect of the level of customer service is due-date performance. A firm that continually has a poor record of on-time shipments will eventually lose market share to competitors that can deliver as promised. Another aspect of customer service is the quoted lead time on new orders. Firms that can promise and deliver on short lead times have a decided advantage over their competitors. Consider, for example, a firm that offers delivery of made-to-order electrical motors in three weeks, while all of their competitors require at least a five-week lead time. To some buyers, the time difference may not matter. But to others, the two-week differential is critical and the business will be awarded to the supplier offering the quickest delivery.

Being the low-cost producer is an obvious competitive advantage. The low-cost producer will have the flexibility to choose between two very desirable alternatives. One alternative is to price the products at a low level in order to capture additional market share. From the buyer's point of view, if all other competitive aspects (such as responsiveness and quality) are equal, then the low-price supplier will normally receive the contract. The other alternative available to the low-cost producer is to keep the price at a level that is on a par with their competitors. This results in a higher profit margin per unit, compared to the competition. Higher profit margins mean a lower break-even point and the potential for increased profits. Large profits, in turn, can lead to opportunities to enhance long-term competitiveness through increased investment in research and development, new technology, employee development programs, and productivity and quality improvement activities.

Synchronized Flows and the Competitive Edge

How can a firm improve product quality, increase customer service, and become a low-cost producer? The answer is by adhering to the basic philosophy of synchronous manufacturing. The validity of this statement can be established through a close examination of the relationship between synchronized product flows and the competitive edge elements.

Synchronized versus Nonsynchronized Flows   It is most illustrative to explore the impact of product flows on the competitive edge elements by analyzing two very different manufacturing environments. Regardless of the various manufacturing control systems that may be used, the only significant distinction between the two environment types is whether or not the product flows are synchronized.

One manufacturing environment, hereafter referred to as the nonsynchronized flow, is characterized by systems where products have long manufacturing lead times and materials spend a large amount of time waiting in queues as work in process. Studies have shown that in most manufacturing plants, the vast majority of manufacturing lead time for materials is actually spent waiting in queues. In some plants, the actual processing time on a given order is as little as 5 percent of total manufacturing lead time. Unfortunately, this scenario is closer to the rule than the exception in most American plants.

The other manufacturing environment is referred to as a synchronized flow. In this environment, products have relatively short manufacturing lead times, and materials spend very little time waiting in queues. Thus, processing accounts for a relatively high percentage of the manufacturing lead time for materials. In order to achieve this result, the flow of materials through the plant must be carefully synchronized, with materials moving smoothly and continuously from one operation to the next.

There is one key point of comparison between these two manufacturing environments. A nonsynchronized flow environment gives rise to high work-in-process inventories and long manufacturing lead times. Conversely, the synchronized flow environment is characterized by low work-in-process inventories and short manufacturing lead times. It will be clearly demonstrated that these two environments have precisely opposite effects on the firm's competitive edge. [5, pp. 36–64]

An Illustration   Suppose that Cherrywood Products, Inc., has received a contract for 80 units of a make-to-order product. This order is normally processed through four different departments (A, B, C, and D), in that sequence. The amount of processing time required in each of the four departments is 1 hour per unit. For simplicity at this point, we assume that any necessary setups are negligible. The plant has one work shift that puts in 8 hours a day, 5 days a week. Thus, each department is available for 40 hours of processing time per week. Figure 2.2 illustrates this situation.

Assume that Cherrywood has the necessary materials to process the order, and the work is ready to begin. Most managers are aware that splitting orders into small batches and overlapping the work can sometimes reduce the manufacturing lead time for the order. Whether or not manufacturing time is actually reduced depends on whether the departments in the process (A, B, C, and D) receive work-in-process materials from other departments outside this particular routing. Nevertheless, in an attempt to expedite the completion of the work, the 80 unit order will be split, overlapped, and scheduled as four batches of 20 units each. That is, all necessary materials are time-released to department A in batches of 20 units every 20 hours. Thus, the last batch of 20 units is released to department A exactly 60 hours after the release of the first batch. The processing of the first batch begins as soon as department A completes any previously received work orders in its queue. Starting with department A, the order is processed and moved in batch sizes of 20 units. Upon completion at department A, the batches are moved to department B where the units are processed. From department B, the product is moved to department C in batch sizes of 20 units. The order is processed and moved in this manner until all work is complete.

Nonsynchronized product flows are the norm in manufacturing environments. Thus, if the product flow at Cherrywood is typical of most plants and is not synchronized, then the following scenario would be quite representative. Assume that because of an unsynchronized work flow and poor scheduling, each department at Cherrywood typically has 2 weeks (80 hours) of work-in-process inventory waiting in queues. In such a system, the 80 unit order, processed by splitting and overlapping four batches of 20 units each, would require 460 hours of manufacturing lead time. To make this determination, let's follow the last batch of 20 through the process. The last

FIGURE 2.2   PROCESSING INFORMATION FOR MAKE-TO-ORDER PRODUCT AT CHERRYWOOD PRODUCTS, INC.

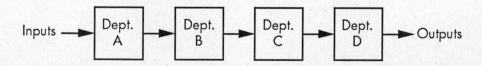

Order Quantity = 80 Units

Processing Time Required per Unit per Department = 1 Hour

Available Time per Week per Department = 40 Hours

batch must wait 60 hours before it is released to department A, where it faces a queue of 80 hours. (Some of the 80 hours of work in process in the queue ahead of the batch includes previously released batches from the same order.) After 140 hours, the last batch begins to be processed at department A and is finished processing at 160 hours. From that point it takes another 80 hours of waiting and 20 hours of processing for each of the following three departments. Thus, the total processing time for the entire order from start to finish is

$$160 \text{ hours} + 100 \text{ hours} + 100 \text{ hours} + 100 \text{ hours} = 460 \text{ hours}.$$

The expected manufacturing lead time for this order is 11.5 weeks. Figure 2.3 illustrates this situation, showing the expected schedule of waiting time and processing time for the 80 unit order at each department.

FIGURE 2.3   **EXPECTED SCHEDULE OF WAITING AND PROCESSING TIME FOR A NONSYNCHRONIZED FLOW ENVIRONMENT AT CHERRYWOOD PRODUCTS, INC.**

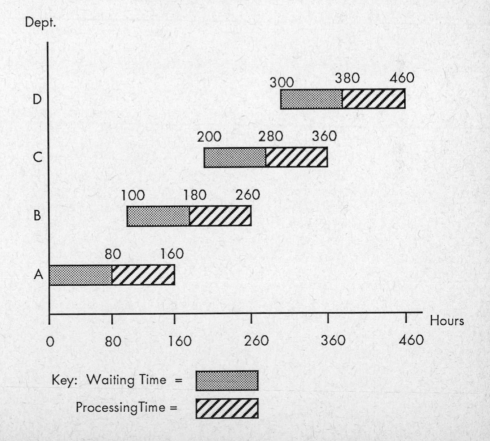

If the flow at Cherrywood is synchronized, then a very different scenario unfolds. A perfectly synchronized flow implies that all materials flow smoothly and continuously through the process without interruption. This means that as a batch arrives at each department, there is no queue, and processing at that department begins immediately. In such a system, the total manufacturing lead time for the 80 unit order is 140 hours. To demonstrate this lead time calculation, let's again follow the progress of the last batch of 20 units as it is processed. The last batch is released to department A and processing begins 60 hours after the release of the first batch. Twenty hours later, this last batch completes processing at A. The batch is moved in turn to departments B, C, and D where processing immediately starts. Processing at each additional department takes 20 hours with no waiting in queue. Thus, the total elapsed time from the release of the order to the completion of the last batch is

$$60 \; hours + (20 \; hours \times 4 \; departments) = 140 \; hours.$$

The expected manufacturing lead time is only 3.5 weeks. Figure 2.4 illustrates this situation, showing the expected schedule of waiting time (none) and processing time for the 80 unit order at each department.

Impact of Product Flows on the Competitive Edge Factors

It is clear that the synchronized and nonsynchronized flow processes represent two very different manufacturing environments. We now examine the effect of these two different environments on the factors of competitive edge.

Quality   It is no secret that the Japanese are the acknowledged leaders when it comes to producing high-quality products. But it hasn't always been that way. Shortly after World War II, Dr. W. Edwards Deming went to Japan to help them embark on their journey to improved productivity and quality. Deming is now almost universally recognized as the foremost quality expert in the world.

Deming's philosophy on quality is not complex. On the contrary, his overall approach to quality is based on 14 fairly simple principles. But the underlying theme of his philosophy is that actions taken in the name of quality should focus on improving the process. Dr. Deming insists that quality cannot be inspected into a product, it must be built in. [12] Following the Deming philosophy, when the Japanese discover a defect, the situation is not treated as a disaster, but rather as another opportunity to improve the process. As a result, an interesting chain of events is set into motion. First, in order to avoid cover-ups, workers are not reprimanded when a mistake is made.

FIGURE 2.4   EXPECTED SCHEDULE OF PROCESSING TIME FOR A SYNCHRONIZED FLOW ENVIRONMENT AT CHERRYWOOD PRODUCTS, INC.

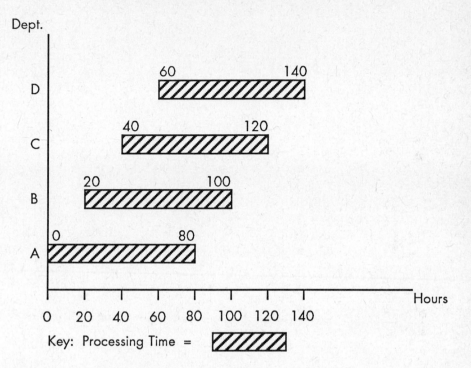

Then, with the assistance of the workers, the cause of the defect is tracked down. Finally, corrective actions are carefully evaluated and implemented in order to reduce the likelihood that the problem will recur.

A critical question is whether or not the manufacturing lead time in an operation affects the ability of management and labor to implement the Deming approach to improving quality. The answer is a thunderous yes!

Suppose that while processing the first unit of the 80 unit order, a piece of equipment in department A malfunctions and begins to produce defective units. The problem is of such a nature that the defect is not detected until the product begins processing at department C. What are the consequences of this problem?

In the nonsynchronized flow environment described above, the problem is not detected until the product reaches department C, 280 hours (7 full weeks) into the process. In the meantime, all of the 80 units are defective and have completed processing at both departments A and B. If the defective units must be scrapped, then the quality problem causes a real catastrophe. If only rework is required, then the problem is probably less disastrous, yet still severe. But in either case, the most pressing problem for management is likely to be how to get the necessary parts expedited or how best to rework the product in order to get the order shipped. Since 100 percent of the order is defective and 280 hours of lead time have passed, this is a sizeable challenge.

The process of identifying the cause of the problem and finding ways to prevent the problem from recurring takes a back seat to getting the order shipped. Moreover, the malfunction that caused the problem in department A took place during weeks 3 and 4, 3 full weeks earlier. What are the chances that the cause of the problem will ever be found? Who can recall the exact operating conditions that existed at department A, which is no longer processing the order? And who would want to admit to being partly responsible for such a grievous error anyway? As a result, it is likely that the cause of the problem will not be discovered, and the process will not be improved. In all probability, the problem will pop up again, and according to Murphy's Law, no doubt at the least opportune time.

If the same malfunction occurs in the synchronized flow environment, when the first batch reaches department C (after 40 hours), department A is preparing to work on unit number 41, starting the third batch. In this situation, only 40 defective units have been produced. The amount of scrap or rework is much less than under the nonsynchronized flow environment.

More important, since department A is still in the process of producing the defective units, the malfunction should be relatively easy to identify and correct. There is no need to speculate on the cause of the problem. Simple observation and checking of the process is likely to reveal the problem. Once the source of the problem is identified at department A, the problem can be corrected, and the process can continue with little loss of time or material. What is just as important, however, is that management can take steps to identify the root cause of the malfunction. Then the control system can be modified so that such a malfunction does not recur without being recognized. The final result is that the process will be improved.

In the synchronized flow environment, not only is the loss of current throughput minimized, but future throughput is protected from a recurrence of the same problem. This illustration clearly shows that it is much easier to produce a high-quality product and achieve a competitive advantage in a synchronized flow environment than in a nonsynchronized flow environment.

It is incorrectly believed by some that cutting the batch size is the key to improving a firm's competitive edge. This is not true in the case of improving quality, nor is it true with respect to the other factors of competitive edge. Although batch size is an important consideration, the key to improving quality and the other factors of competitive edge lies in synchronizing the product flow through the plant. The Cherrywood illustration can be used to clearly demonstrate this point.

In a synchronized flow environment, reducing the batch size does improve the degree to which the product moves quickly through the process in a smooth and continuous manner. Consider a perfectly synchronized environment, with a batch size of 1. If there are no disruptions or problems in the process, the 80 unit order will be completely processed in 83 hours. If there is a malfunction at department A, then the first defective unit reaches department C after only 2 hours. Thus, the malfunction problem is discovered almost immediately, and only 2 defective units would be produced before the problem is recognized. Thus, in the case of a plant where the flow is already synchronized, reductions in the batch size can result in further improvements in the firm's competitive performance.

In the nonsynchronized flow environment, even if the batch size is cut to 1 unit, the total manufacturing lead time for the order is 403 hours. Furthermore, any malfunction at department A that is discovered at department C would not take place until 242 hours have passed. By that time, all 80 units would have passed through department A and would be defective. The last defective unit would have been processed through department A 2 weeks earlier. Thus, cutting the batch size in the nonsynchronized flow process is of very limited benefit. The only real difference is that department A will complete the processing of the order 2 weeks prior to discovery of the problem instead of the previous 3 weeks.

Customer Service   The manufacturing lead time for any order is highly dependent upon a number of variables. But the critical ingredient that establishes a general level of manufacturing lead time within any operation is the degree to which the material flows are synchronized. In general, the greater the degree of synchronization, the lower the level of work-in-process inventory, and as a result the shorter is the manufacturing lead time.

The Cherrywood illustration can be used to demonstrate a basic principle at work here. Consider the effect of doubling the work-in-process inventory in the nonsynchronized flow environment. That is, at each department, suppose the queue is now at 160 hours of work instead of the previous 80 hours. Using the original batch size of 20 units, the resulting manufacturing lead time increases from 460 to 780 hours.

The principle at work in this illustration is very simple. The manufacturing lead time for any order (assuming the order is not given any special priority) is roughly proportional to the amount of work-in-process inventory queued up in front of that order. The greater the level of work-in-process inventory that is carried within any system, the longer the manufacturing lead time for that system. (It should be understood that the degree to which the relationship is truly proportional is dependent upon the relative size of total processing time compared to total queue time.)

The bottom-line result is clear. Those firms that carry excess work-in-process inventory to protect the process and keep individual efficiencies high are guaranteeing that their system will have an artificially high manufacturing lead time. The undeniable result is a system that is less responsive to the market. This unresponsiveness might be very costly in terms of market competitiveness.

To illustrate, suppose that the standard lead time required by market competitors to produce products similar to Cherrywood's product line is generally about 7 weeks. Since buyers are normally familiar with the market lead time for products, most buyers would typically place their purchase orders about 7 to 10 weeks before the order is needed. (Although many vendors will swear that some buyers never plan that far into the future.) If Cherrywood (operating a nonsynchronized flow process) has a quoted lead time of 11.5 weeks while its competitors have a quoted lead time of 7 weeks, who is most likely to get orders? Conversely, what if Cherrywood (operating a synchronized flow process) has a quoted lead time of 4 weeks based on its actual manufacturing lead time of 3.5 weeks? Clearly, Cherrywood now has the inside track on landing new orders. This is especially true for customers that require a lead time of less than 7 weeks.

From the perspective of gaining the competitive edge in the marketplace, it is not necessary that manufacturing lead times be zero. It is sufficient that a firm's lead time be less than their competitors' lead times.

Another important aspect of customer service is due-date performance. Not surprisingly, a firm's due-date performance is also a function of the manufacturing lead time. But another concept—the validity of the product forecast—plays an important role in understanding due-date performance.

Every firm works off a product forecast, whether it be formal or informal. And that forecast is usually fairly accurate for some period of time into the future. But then the reliability of the forecast degenerates quite rapidly. The primary factor that determines the time horizon for which the forecast has high validity is the standard quoted lead time in the market. If a firm has a manufacturing lead time that is less than the standard lead time for the market, then that firm has the advantage of being able to start production based on a product forecast that has high validity. Conversely, if the firm has a manufacturing lead time that exceeds the market lead time, the product forecast will be of low validity.

Figure 2.5 illustrates how the forecast validity for a firm changes over time. The figure also indicates that the standard quoted lead time in the market is 7 weeks. Notice that the validity of the forecast drops rapidly after 7 weeks. This is because almost all future orders that are required 7 weeks into the future have already been received, and these orders are normally considered to be firm orders. Once past the 7 week mark, however, the number of firm orders decreases. And there is less reliability the further out into the future that orders are forecasted. Even those buyers that place orders several months in advance will sometimes feel free to change their order anytime prior to the last 7 weeks, since that is the normal lead time for the market.

Figure 2.5 also indicates the relative degree of forecast validity for 3.5 weeks versus 11.5 weeks of manufacturing lead time. In the synchronized

FIGURE 2.5 THE DETERIORATION OF PRODUCT FORECAST VALIDITY OVER TIME

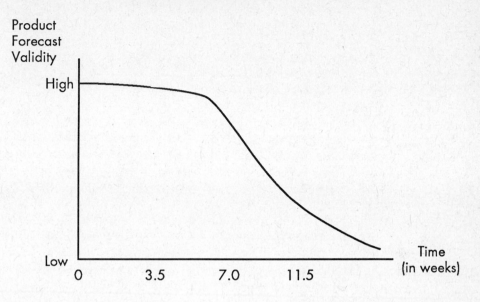

flow environment, where the manufacturing lead time is 3.5 weeks, the forecast validity is extremely high. That is, production starts, based on highly reliable information concerning firm orders. Conversely, in the nonsynchronized flow environment, with a manufacturing lead time of 11.5 weeks, the forecast validity is extremely low. What if the market requires delivery in 7 weeks from receipt of the order? In order to meet due dates, production would have to begin 4.5 weeks in advance of receiving the order. Thus, production starts, based on poor knowledge of actual product demand. As a result, the firm will most likely be plagued with missed due dates as well as large work-in-process and finished goods inventories for which there is no demand.

Product Cost   In order to be the low-cost producer in the market, a firm must effectively manage all aspects of the operation. In this section, it is demonstrated that the extent to which product flows are synchronized has a huge impact on product cost.

Two variables that bear heavily on the cost of the product are operating expenses and the required level of investment in plant and equipment. In order to become a low-cost producer, a firm must be able to keep operating expenses at a minimum and eliminate expenditures for unnecessary equipment.

In plants with lengthy manufacturing lead times, it seems as if the only way to get orders shipped on time is to expedite. The procedures typically used in such an operating environment require huge sums of money for overtime, premium freight, broken setups, and misallocation of productive resources and materials. There are always unforeseen emergencies, but if this scenario occurs with any regularity, it is a sure sign of deep-seated problems. In an effort to cure the symptoms and protect themselves from these costly practices, managers may come to rely on large stocks of work-in-process inventory. But resorting to such a solution is like throwing gasoline on a fire in an attempt to put it out.

Let's reconsider the order for 80 units at Cherrywood. Suppose that to get the order, the sales representative had to promise delivery in 7 weeks, the market norm. But what if Cherrywood operates in a nonsynchronized flow environment and has a manufacturing lead time of 11.5 weeks? If the normal work schedule is followed as planned without the order receiving special priority, then to finish the order on time, massive amounts of overtime will be necessary. Therefore, to get the order out on time, operating expenses for the 7 week time period will be much higher than planned. The only way to avoid overtime on this particular order is to assign it a special priority. This assigned priority will allow the order to bypass other orders already in the queues, resulting in an earlier completion date. But expediting this one order will result in the displacement of other scheduled work. This displacement will cause other work orders to fall behind schedule, which only compounds the problems for the entire system. The result is a vicious cycle of continual expediting costs. This inflated level of operating expense makes it impossible to be a low-cost producer and cuts into the firm's profit margin.

Operating in a nonsynchronized flow process environment also affects the level of investment in equipment. It is a common occurrence in plants that the last one or two operations in the process have the greatest productive capacity in the entire plant. The explanation for this phenomenon is simple.

In all manufacturing plants, there is always great pressure to meet due dates. But the greatest pressure to ship orders naturally occurs as the due date draws near. At this point, hopefully, the order is usually close to being completed. That is, the order is scheduled for processing at the last one or two operations. All the pressure is therefore focused on the supervisors at these last operations in the process. If the order is behind schedule, then these last operations are expected to make up the lost time and put the order back on schedule. Of course, the cry from the supervisors at those operations is always that more capacity is needed to meet these periodic crises. Furthermore, these operations have high visibility to management since they appear to be the key to meeting due dates and also meeting monthly shipping quotas. The result is that requests to purchase additional or higher-capacity equipment for these last operations are typically approved.

Much of the time, the extra equipment that is purchased for the last operations is grossly underutilized. As a result of the logistics involved in the nonsynchronized flow environment, the investment in equipment is much higher than it should be. This excess investment acts to further increase the cost of the product and reduce the potential profit margin for the product.

In contrast, there is typically plenty of time to meet shipping deadlines in synchronized flow environments. Thus, there is little need for frantic and expensive expediting. In addition, the more synchronized the product flow, the less likely it is that the workload for resources will alternate between overloads and underloads. Therefore, there is no undue pressure to purchase excess equipment. The manufacturer that operates a synchronized flow process is well on the way to becoming a low-cost producer.

The Operational Measures and the Competitive Edge

It is now clear that establishing a synchronized flow within a manufacturing environment is critical to gaining a competitive advantage over market competitors. But it must be recognized that there is no finish line in this endeavor. The target is a moving one. As a firm continues to improve the synchronization of its operation, new opportunities for improvement will present themselves. The truly competitive firm will establish an ongoing process of trying to improve the synchronization of the product flow. The result will be continual improvements in product quality, customer service, and production costs.

As a firm moves toward a synchronized flow environment, improvements in product quality, customer service, and production costs should be evident. But how do these improvements relate to the financial health of the firm? The answer is that changes in the competitive edge factors will eventually show up as changes in the operational measures of throughput, inventory, and operating expense, which in turn can be expressed in bottom-line financial terms. From the opposite perspective, improvements in throughput, inventory, and operating expense will have beneficial effects on the competitive edge factors. Some of the key relationships between the factors of competitive edge and the operational measures are now briefly considered.

Throughput   Throughput is significantly affected by a firm's competitive edge. Without an increase in market demand, any increase in throughput must be temporary. Since market demand is a function of the firm's competitive position in the market, permanent increases in throughput are normally the result of an improvement in the firm's competitive edge.

If a firm is operating at peak capacity and still not satisfying all available market demand, a special situation exists. In this case, quality and process improvements *might* have a significant effect on throughput. For example, reductions in scrap or required rework at critical points in the process may result in increased throughput.

It has already been demonstrated that an increase in throughput positively affects all the bottom-line measures. But another benefit is the spreading of the overhead and operating expense over a larger number of units. This will result in a lower cost per unit produced. Thus, an increase in throughput will actually enhance the firm's competitive position.

Operating Expense   The quality element of competitive edge can positively influence operating expense in one fundamental way. Quality has the same effect on operating expense that it does on throughput. If the quality of a process is improved, it may be possible to reduce the level of operating expense. A good illustration of this possibility is found in plants that operate separate rework lines. An improvement in quality would require less equipment and fewer workers to operate these lines. A high level of quality, such as typically found in Japanese plants, would make such rework lines totally unnecessary. Operating expense would decrease.

On the other hand, the level of operating expense significantly influences the firm's competitive edge. Reductions in operating expense will reduce the cost per product and make the firm more competitive. However, a negative side also exists. Cuts in operating expense might adversely affect the quality of the product being produced. A typical and potentially more serious scenario occurs when cuts in processing capacities are made within the plant. These cuts are made in the name of reducing operating expense and are expected to help balance the plant. What often happens, however, is that these reductions in capacity are implemented to the point that throughput is threatened. The negative implications of trying to balance the plant are discussed in greater detail in Chapter 3.

Inventory   Just as the quality level for a process affects throughput and operating expense, it also affects inventory. A reduced level of scrap and rework throughout the plant will improve the overall functioning of the operation and result in a lower dependence on "just-in-case" inventories. But this relationship is not the one that needs to be emphasized.

It was hinted earlier in this chapter that the impact of inventory in the manufacturing environment is quite often misunderstood and underestimated by managers. The effect of plantwide, work-in-process inventory levels on the competitiveness of the firm may be the most significant and most overlooked relationship within the manufacturing environment.

Most managers are aware of the traditional costs associated with carrying inventory. But the traditional cost to carry inventory is really insignificant in comparison to the loss of competitive advantage that occurs in the high work-in-process inventory environments characterized by nonsynchronized flow processes. The manufacturer that operates in such a high work-in-process inventory environment will continually struggle to survive in today's competitive market.

SUMMARY

Traditional cost accounting systems provide the basis for most of the key decisions made in manufacturing firms in the United States. But this approach has some fundamental flaws. It focuses too much attention on direct labor and ignores the effects of interactions that exist in manufacturing organizations. This has resulted in a highly localized managerial perspective that overemphasizes the reduction of direct labor. The typical result is a highly nonsynchronized flow of products through the plant that adversely affects the firm's ability to compete.

A firm's competitive edge is based on its ability to produce high-quality products, provide superior delivery performance, and maintain a competitive product cost. To achieve these aspects of competitive edge, a firm must be able to establish a synchronized flow of products through the plant that matches the marketplace demand in a timely manner.

The operational measures of throughput, inventory, and operating expense can help managers develop a more effective global perspective for decisions affecting the firm's operation. Focusing on these measures will help managers evaluate the impact of specific manufacturing actions and decisions on overall system performance. This, in turn, will enhance the quality of decision making throughout the organization and improve both the competitive position and profitability of the firm.

QUESTIONS

1. Identify some of the major subsystems of a manufacturing firm.
2. What are the basic assumptions of the standard cost system?
3. State the assumption of reversibility. What is its role in the decision-making aspect of the standard cost system?
4. How does the assumption of reversibility simplify management's job?
5. Why are the cost savings from new and faster machines seldom realized?
6. Define the operational measures. How are they related to the bottom line of making money?
7. List the competitive elements by which manufacturing firms compete.
8. Compare synchronous and nonsynchronous manufacturing flows with respect to a) quality, b) responsiveness, c) due-date performance, and d) product cost.
9. Discuss the fundamental changes taking place in the marketplace and in manufacturing processes that have made it increasingly difficult to remain profitable without achieving a synchronized product flow.
10. Discuss the relationship between forecast accuracy and synchronization of flows.

PROBLEMS

1. An engineer in a manufacturing plant proposes a process change that is designed to improve the yield from 80% to 90%. Use the information given to calculate the expected savings and payback period.

   > Market demand is approximately 1,000 good units per year.
   > Processing time using the current method = 1 hour.
   > Processing time using the proposed method = 1.1 hours.
   > Cost of direct labor = $15 per hour.
   > Cost of material per unit = $50.
   > Overhead factor for this process = 2.20.
   > Cost of implementing the new process = $5,000.

2. These numbers describe a manufacturing company.

   > Net revenue = $120,645,122.
   > Cost of goods = $80,314,210.
   > Cost of goods breakdown: Labor = 9%, Material = 37%, Overhead = 54%.
   > Inventory = $26,122,300.
   > Inventory breakdown: Material = 48%, Value-added = 52%.
   > Selling, administration, and other expenses = $24,989,673.
   > Total assets = $84,229,105.
   > Inventory carrying cost = 13 percent per year.

   a. What are the synchronous manufacturing values of throughput, inventory, and operating expense for this company?
   b. Assume that throughput increases by 10% and inventory and operating expense remain the same. What is the impact on net profit, return on assets, and cash flow?
   c. Assume that operating expenses decrease by 5% and throughput and inventory remain the same. What is the impact on net profit, return on assets, and cash flow?
   d. Assume that inventory decreases by 5% and throughput and inventory remain the same. What is the impact on net profit, return on assets, and cash flow?
   e. What is the impact if throughput increases by 10% at the same time that operating expense and inventory both decrease by 5%?

Resource and Product Interactions

INTRODUCTION

The standard cost system does not take into account the numerous interactions that exist between resources and products. As discussed in the last chapter, this is one of the major weaknesses of the standard cost system. Ignoring resource/product interactions inevitably leads to poor managerial and operating decisions. In order to improve our decision-making ability, it is necessary to understand the significance of these interactions in the manufacturing environment. In this chapter, the origin of these interactions and their impact on the manufacturing operation are examined in detail.

THE BASIC PHENOMENA OF MANUFACTURING

Every manager is aware that it is very difficult to achieve a smooth and continuous flow of products in concert with market demand. What is not generally recognized is that the major obstacles to running an effective operation are the result of two basic phenomena of manufacturing. [31] One basic phenomenon of manufacturing is the existence of dependent events and the resulting interactions between resources and products. The other basic phenomenon is the occurrence of statistical fluctuations and random events within every manufacturing environment.

Dependent Events and Interactions

Manufacturing operations are characterized by the existence of numerous dependent events and interactions. By dependent events, we mean that certain operations or activities in the plant cannot take place until certain other operations or activities occur. The term interaction has a slightly different connotation than dependent events; interaction would normally be used to indicate the effect that dependent events have upon one another. In this text, both terms will be used to demonstrate the dependencies that exist within manufacturing environments.

The routing sequence of required operations for the production of a product is a simple example of manufacturing dependency. In the typical manufacturing environment, the production process does not begin until the required materials have been procured. Individual operations in the sequence are not performed until the previous operations specified in the routing have already been completed. And assembly cannot be completed until all components have been purchased and/or fabricated.

A different but equally important type of dependency comes into play when one resource is required to process two or more different products. When two or more products are processed at a resource, scheduling becomes an important issue. Some of the most perplexing operating decisions in a manufacturing firm involve scheduling problems. Every manager understands that a poor schedule at almost any resource can adversely impact both the validity of schedules at other resources and the product flow. But the impact goes much deeper than that. The resultant problems can very negatively affect throughput, inventory, and operating expense.

It will be demonstrated that dependent events, and the inevitable resource and product interactions, are an integral part of manufacturing operations. Moreover, the successful management of a manufacturing operation requires that special attention be given to the role played by these dependencies.

Statistical Fluctuations and Random Events

The significance of dependent events and interactions in the manufacturing process is magnified by another inescapable reality of manufacturing systems. That reality is the universally unavoidable existence of random events and statistical fluctuations within a manufacturing environment.

Random events are those activities that take place at irregular intervals and have a disruptive effect on the manufacturing process. A major cause of concern over random events is that they cannot be predicted with any degree of reliability. Random events can be introduced into the process from many sources, and the element of uncertainty and disruptiveness caused by random events can never be completely eliminated. The availability of materials, tools, and skilled labor to perform a given process are all subject to random disruptions. Machines do not work forever without breaking down or needing adjustment, tools break or are misplaced, and workers get sick. Furthermore, these problems can occur anywhere and at any time. The only rule they seem to follow is one of the corollaries to Murphy's Law—disruptions will occur at the least opportune time and at the most expensive spot!

In manufacturing environments, the term *statistical fluctuation* refers to the concept that all processes have some degree of inherent variability. While operations that show statistical fluctuations may be more predictable than random events, they still introduce a great deal of instability into the process. On the purchasing end, the delivery of material from even the most reliable of vendors is subject to a certain level of uncertainty. On the production floor, these fluctuations seem to be rampant. The time required to set up a resource for a particular operation, the time required to process a given product at a given resource, and the time required to move the product to the next operation are all subject to statistical fluctuations. But whenever people are heavily involved in the process, the amount of time required to complete a task will be especially variable. Not surprisingly, these fluctuations are not confined to the factory floor and vendors. The market is subject to similar fluctuations in the form of demand changes from the customer and/or changes in the sales forecast.

It is apparent that random events and statistical fluctuations describe slightly different phenomena. However, over the long run, even random events generally exhibit some statistical properties of recurrence. Therefore, the single term—*statistical fluctuations*—will normally be used to describe the variability that exists in manufacturing processes.

It should be evident that these two phenomena, namely (1) dependent events and interactions, and (2) statistical fluctuations and random events, combine to wreak havoc upon ill-conceived and poorly controlled manufacturing management systems. In fact, the day-to-day managerial role of a shop floor supervisor is often nothing more than attempting to cope with the almost endless stream of disruptions and adjustments to planned activities.

IMPACT OF THE BASIC PHENOMENA ON MANUFACTURING OPERATIONS

To demonstrate the destructive effects of the phenomena of dependent events and statistical fluctuations in manufacturing systems, it is useful to construct an analogy. Several such analogies are possible. In this section, a column of soldiers on a forced march is utilized to illustrate the impact of these phenomena on manufacturing operations. (A more complex representation of the production process can be demonstrated through the use of a production dice game. In this approach, dice are used to simulate statistical fluctuations occurring in a production process. The dice game and some of its variations are discussed in the Appendix.)

A Useful Analogy

A column of soldiers on a forced march is analogous to a plant when several transformations are made. [31] In order to fully develop this analogy, the following procedure will be followed. First, a simple process in a manufacturing plant is described. Second, the characteristics of a troop of soldiers on a forced march are described in terms of resources, inputs, and outputs. Third, the manufacturing concepts of throughput, inventory, and operating expense are illustrated in terms of the analogy. Finally, the analogy is used to illustrate some very important principles of manufacturing processes.

A Simple Production Process   Imagine a simple manufacturing process that produces only one product, uses only one raw material input, and has only one possible sequence of operations. The material is received and processed by the first resource. It is then transferred to the second resource where it is processed and transferred to the third resource. This procedure is continued until the material has been completely processed by all of the resources (a total of n resources) in their proper sequence. Only after the material has been processed by all of the resources is there a finished product. Clearly, an order is not completed until all units of the product have been processed by all of the resources. This simple process is illustrated in Figure 3.1.

A Column of Marching Soldiers   The objective of a forced march is to cover a certain amount of ground in a limited amount of time. The raw material that is being processed is the ground over which the soldiers march. The resources are the soldiers that must cover the ground. The routing sheet consists of one operation (marching) for every row of soldiers and requires that the ground is covered by one row of soldiers at a time. Each row of soldiers must cover the ground in sequence. The first row of soldiers must cover a piece of ground before the second row, and the second row must cover a piece of ground before the third row, and so on. The finished product

FIGURE 3.1   A SIMPLE STRAIGHT-LINE PRODUCTION PROCESS

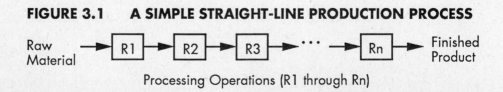

Processing Operations (R1 through Rn)

is the ground that has been covered by the entire troop. The march is completed only when every soldier in the troop has traversed all of the designated ground. The basic characteristics of the forced march are illustrated in Figure 3.2.

Parallels between the March and the Production Process　There are some obvious parallels between the forced march and the simple production process described. The fundamental considerations of raw material, resources, processing sequence, finished goods, and the concept of a completed objective have been described thus far. But there are several additional fundamental parallels that are now identified.

In the production process, any material that has been processed by one or more, but not all, of the resources is termed *work-in-process inventory*. In the forced march, any stretch of ground that has been covered by some, but not all, of the soldiers is partially processed material and therefore represents work-in-process inventory.

In the production process, the manufacturing lead time is the length of time between the release of materials into the system and the transformation of those materials into a finished product. In the forced march, lead time is the time interval between the first and last row of soldiers crossing a specified point of ground.

In the production process, throughput is the amount of product produced and sold. If it is assumed that all goods produced are immediately sold, then the throughput rate is the actual production rate of the last resource in the process. In the forced march, throughput is represented by the amount of

FIGURE 3.2　THE FORCED MARCH ANALOGY

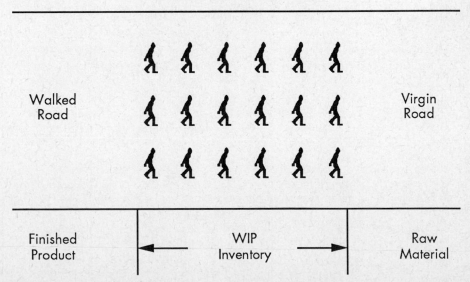

| Walked Road | | Virgin Road |

| Finished Product | WIP Inventory | Raw Material |

ground covered by the entire column of soldiers. The throughput rate is the speed of the last row of soldiers.

In the production process, operating expense is the cost incurred in turning materials into throughput. In the forced march, operating expense is represented by the amount of energy expended by the soldiers to achieve their objective.

The parallels between the forced march and the simple production process described here are summarized in Table 3.1.

TABLE 3.1   PARALLELS BETWEEN THE FORCED MARCH AND THE SIMPLE STRAIGHT-LINE PRODUCTION PROCESS

SIMPLE PRODUCTION PROCESS	FORCED MARCH
Raw materials are to be processed	Ground is to be covered
Resources are used to process materials	Soldiers are used to march on the ground
Material is processed by one resource at a time	Ground is covered by one row of soldiers at a time
Work-in-process inventory is the amount of material between the first and last resource	Work-in-process is the amount of ground between the first and last row of soldiers
Manufacturing lead time is the period of time from the release of material into the system until the material is processed by the last resource	Lead time is the period of time from when the first row of soldiers marches over a specific piece of ground until the last row marches over that same ground
Throughput represents the amount of product produced and sold.	Throughput represents the amount of ground covered by the entire column of soldiers
Throughput rate is the amount of product produced by the last resource and sold by the firm	Throughput is the amount of ground covered by the last row of soldiers (not to exceed the objective of the march)
The objective of the production process is to produce a given amount of throughput in a given period of time	The objective of the forced march is to cover an assigned amount of ground in a given period of time
Operating expense is the cost incurred by turning materials into the desired throughput	Operating expense is the amount of energy expended by the soldiers in completing the march

Effect of the Basic Phenomena
on the Operational Measures

Consistent with the approach developed in Chapter 2, the impact of dependent events and statistical fluctuations on the manufacturing environment will be examined in terms of the operational measures of throughput, inventory, and operating expense. This approach allows us to analyze the systemwide effects that these two phenomena have on manufacturing operations. The forced-march analogy provides the vehicle by which this analysis is conducted.

In the forced-march analogy, a significant dependency was identified. The dependency was the limitation that a row of soldiers could not pass another row of soldiers. Thus, the marching speed of any given row of soldiers cannot exceed the marching speed of the row immediately ahead. This creates a chain of dependencies throughout the entire column. Ultimately, each soldier in the column plays a significant role in determining whether the column achieves its objective in a timely manner. This dependency is similar to that which exists in the simple production process where each resource has the potential to affect adversely the timely completion of scheduled throughput.

Statistical fluctuations also exist in our forced-march analogy. The rate at which the soldiers march is analogous to the rate at which manufacturing operations are performed. Obviously, all soldiers do not inherently have identical marching paces and strides. Furthermore, the rate at which any given soldier marches exhibits random fluctuations. A soldier's stride is not perfectly constant all the time, and he may even occasionally stumble. Likewise, in a manufacturing plant, each resource has a different processing capacity that exhibits statistical variation from the average rate.

To understand the impact of statistical fluctuations (the variations in the marching speed of the soldiers over a period of time) and the dependent events (one row of soldiers cannot pass a row of soldiers in front of them), consider what would happen to a column of soldiers as they set out on a march. As shown in Figure 3.2, at the start of the march the rows are compact and evenly spaced. But anyone who has participated in such marches knows what happens as the march progresses. As time goes by, gaps begin to form between the lines. Gaps appear and may widen whenever any row of soldiers is moving slower than the row immediately in front of it. When two rows are marching in close formation and the first row slows down, the second row must slow down as well. But when the first row speeds up (thereby attempting to maintain their average marching speed), the second row may not.

The fluctuations that affect the different rows of soldiers occur randomly and are independent of one another. But the positive and negative fluctuations do not completely cancel each other. When a positive fluctuation (higher than average marching speed) occurs, the net effect on the troop is usually limited because of two factors. First, one row of soldiers cannot pass the row immediately in front of them. Second, even if one row of soldiers is able to speed up, the next row may want to speed up also, but may be unable to do so because of their own limitations. Thus, the potential gain in speed is unlikely to spread through the troop. However, whenever a negative fluctuation (slower than average marching speed) occurs in a single row, it has the potential to affect every following row of soldiers. As an extreme example, if one row of soldiers comes to a complete stop, every following row must also eventually come to a halt. The net result is that the negative fluctuations tend to be cumulative in nature and are only somewhat offset by the corresponding positive fluctuations. As a result, in a column of marching soldiers the gaps between ranks grow, and the column continues to spread out over time. This spreading phenomenon is illustrated in Figure 3.3.

The spreading of a column of marching soldiers is the result of the combination of dependent events and statistical fluctuations. If the soldiers were robots with zero variation (no fluctuation) in stride and rhythm and had no breakdowns, then the column would not spread out. On the other hand, if the rows could intermingle (no dependency) then one row could not cause another to slow down; nor could one row prevent another from speeding up. Each row would be marching independently, and the spreading

FIGURE 3.3   THE SPREADING OF THE TROOPS IN THE FORCED MARCH

After 5 Miles

After 10 Miles

would not accumulate. Thus, if either of the two phenomena is removed, the spreading effect will essentially disappear.

Now consider what happens with both dependent events and statistical fluctuations present in a manufacturing system. The natural result is the manufacturing equivalent of the spreading of the column, and the consequences are very serious. In the forced-march analogy, as the column spreads, the distance between the first row and last row of soldiers grows. This increased distance represents an expansion of inventory in the system. In a manufacturing plant, raw material is released into the system at the scheduled rate. But downstream resources fall further and further behind the work schedule as the disruptions and negative fluctuations accumulate throughout the process. One obvious result is an increase in the work-in-process inventory. But there is an even more serious consequence. In the forced-march analogy, the last row of soldiers has covered less distance at the end of the scheduled time period than had been planned. Throughput is less than expected. The increase in the length of the troop column indicates how far short the entire column is from achieving the scheduled throughput. Likewise, in the manufacturing environment, the increased amount of work-in-process inventory translates roughly into lost throughput.

As long as the soldiers have the ability to run, the spreading can be controlled. Every time a significant gap begins to form, the commanders in charge of the column can prompt the soldiers to close ranks by running. This is exactly the expediting action used by all military sergeants. The necessity for the soldiers in the rear rows to make up for the accumulated deviations of all rows ahead of them means that these rows experience the run-and-stop syndrome. More energy is expended by these soldiers than the soldiers in the first row. The soldiers in the first row have to compensate only for their own actions and this they do without any extra effort. This accounts for the desire of all experienced soldiers to be in the front of a column when required to participate in extended marches. (Of course, war fronts introduce different priorities, and the optimal position within the column changes.)

Running to close gaps in the column of soldiers is analogous to increasing the rate of production at specific resources in a manufacturing plant. However, increasing the rate of production of a manufacturing resource is usually not accomplished by simply working faster. Instead, work-in-process buildups are reduced by working more hours. The manufacturing system equivalent of running to close ranks is the use of overtime.

What has been demonstrated through the forced-march analogy is that dependent events and statistical fluctuations have serious consequences on the smooth operation of manufacturing plants. Negative variances in the product flow will accumulate more readily than the positive variances. Thus, the disruptions will not average out for the total system, and most individual resources will perform below their capability. The ultimate result is that the normal flow of products is disrupted to the extent that throughput is lost, excess work-in-process inventory is created, and operating expenses are inflated.

It will also be demonstrated that the impact of dependent events and statistical fluctuations has important implications concerning the capacity of the manufacturing resources. Due to the accumulation of deviations in all manufacturing operations, it will not be sufficient to carry just the right capacity at all resources (as measured by their average performance). This concept is the subject of the next section.

The Attempt to Balance Capacity

Consider a column of soldiers on a forced march where each soldier has the same average marching speed of 3 miles per hour, and the troop is required to cover a distance of 30 miles in 10 hours. If the soldiers' ability to close gaps is taken away (no running allowed), then there is no way to prevent the spreading of the column from becoming excessive, and the march will be completed behind schedule. Now consider the analogous manufacturing operation in which every resource has just enough capacity to meet market demand. If this balanced plant has no reserve capacity (such as overtime), then the existence of dependent events and statistical fluctuations will cause work-in-process inventories to increase, and throughput will fall short of market demand levels.

The Effect of Disruptions in a Balanced Plant The effect of disruptions in a plant with balanced resource capacities can be easily demonstrated. Figure 3.4 shows a simple manufacturing operation that involves one product and two processes. Each process requires an average of 8 hours to complete a job. The first process is performed by resource R1, a conventional machine with a performance distribution shown in Figure 3.5(a). The second process is completed by resource R2, a machine-controlled process that shows very little deviation in its performance. The performance distribution curve for R2 is shown in Figure 3.5(b).

FIGURE 3.4   A MANUFACTURING OPERATION WITH ONE PRODUCT AND TWO PROCESSES

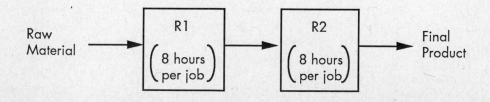

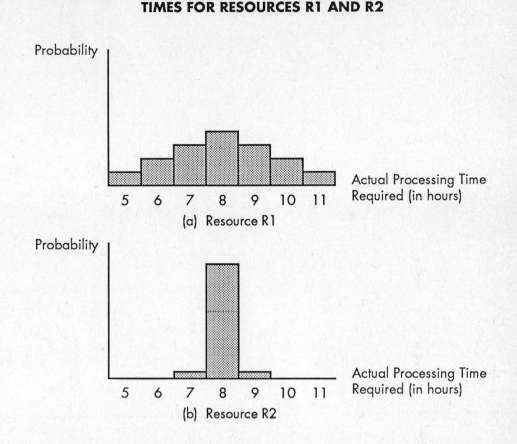

(a) Resource R1

(b) Resource R2

A schedule for the next five jobs to be processed through this two-step operation over the course of 6 working days is shown in Table 3.2. Table 3.3 shows a simulated performance of the five jobs against the planned schedule. The simulation indicates that both resources have performed according to their average capability. There were no extraordinary variances. Both resources have processed five batches of products in 40 hours of machine time, thus meeting the process standard. Nevertheless, at the end of 6 days, the plant is unable to ship the products as expected. Despite the fact that both resources have performed to standard and resource R1 has completed all five batches per schedule, the plant is 4 hours behind schedule.

In this illustration, it is difficult to determine whether the problem is with resource R1 or resource R2. The difficulty arises because the problem is neither resource by itself. The problem is in the interaction between the resources. In this extremely simple example, the product provides the primary medium for the interaction. The original disruption or delay occurs at resource R1. But the product carries this disruption to resource R2, which then falls

TABLE 3.2   SCHEDULE OF FIVE JOBS THROUGH THE SIMPLE TWO-PROCESS OPERATION

JOB NUMBER	RESOURCE R1 Scheduled Start and Finish Times (in hours)	RESOURCE R2 Scheduled Start and Finish Times (in hours)
1	00 - 08	08 - 16
2	08 - 16	16 - 24
3	16 - 24	24 - 32
4	24 - 32	32 - 40
5	32 - 40	40 - 48

TABLE 3.3   COMPARISON OF THE SCHEDULED AND ACTUAL COMPLETION OF THE FIVE JOBS THROUGH THE SIMPLE TWO-STEP OPERATION

JOB	RESOURCE R1 Scheduled Start and Finish Times (hours)	Actual Start and Finish Times (hours)	Deviation (Cumulative)	RESOURCE R2 Scheduled Start and Finish Times (hours)	Actual Start and Finish Times (hours)	Deviation (Cumulative)
1	00 - 08	00 - 10	+2	08 - 16	10 - 18	+2
2	08 - 16	10 - 20	+4	16 - 24	20 - 28	+4
3	16 - 24	20 - 28	+4	24 - 32	28 - 36	+4
4	24 - 32	28 - 34	+2	32 - 40	36 - 44	+4
5	32 - 40	34 - 40	0	40 - 48	44 - 52	+4

behind schedule. If resource R2 were processing several different products (P1 and P2), then any delay experienced in finishing product P1 would carry over to product P2. Product P2 would then become the carrier of this disruption to other resources that are required to process it.

The principle at work here is very significant. Products will carry disruptions from one resource to another, while resources will carry disruptions from one product to the other. Due to the propagation of disruptions along both the product and resource dimensions, any significant disruption or fluctuation will spread through the plant very rapidly. This spreading of disruptions is illustrated by Figure 3.6. In this figure, two parts, A and B, are being processed. A segment of the routing for each part is shown, part A passing through resources R1 and R2, and part B passing through resources R1 and R3. Let's assume that part A is scheduled to be processed first, followed by part B. Suppose that a machine at R1 breaks down while working on part A. This disruption at R1 causes the completion of part A to be delayed. Since part A will be late in arriving at R2, this also causes R2 to fall behind schedule. In addition, the disruption at R1 also affects the timely processing of part B. And since part B becomes delayed, the disruption is now carried to R3 as well.

This illustration demonstrates that the dependencies in the typical manufacturing environment are much more complex than those identified in the forced march. Therefore, it appears that the possible consequences of dependent events and statistical fluctuations are more severe in a manufacturing plant than in the forced march.

The negative effects of having a column of soldiers with no catch-up capability is fairly obvious. The consequences of a perfectly balanced manufacturing plant, with no excess capacity, are not as well understood. In a manufacturing plant, excess or unused capacity at a given resource is usually translated to mean excess cost. In the traditional cost-driven system, this excess cost is often the focus of cost-reduction projects. In a vain attempt to minimize the cost at each process/resource, many manufacturing managers spend considerable time trying to balance the resource capacities in their plants. But as excess capacity is eliminated, the catch-up capability of the various resources disappears and the inevitable happens. The plant begins to fall further and further behind the production plan (gaps will form and grow) as work-in-process inventories increase and throughput lags. The blame for falling behind schedule is attributed to uncontrollable factors. Meanwhile, managers must resort to the use of overtime or other available means of increasing the capacity, in order to meet the production plan. Ironically, managers are soon paying a premium price for the capacity they previously worked so hard to trim.

FIGURE 3.6   DISRUPTIONS SPREAD THROUGH PRODUCTS AND RESOURCES

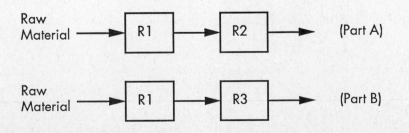

The previous discussion is not meant to be an argument for having great amounts of excess capacities. It is meant to emphasize that the focus on individual resource capacities encouraged by traditional cost systems is ill conceived. The consequences of ignoring the effects of the basic phenomena of manufacturing are potentially devastating. Managers must therefore learn how to operate more effectively within the complex realities of the manufacturing environment.

Constraints on Balancing Capacity In reality, the capacities of individual resources in any manufacturing operation are not balanced. This is the result of two facts: (1) Capacity comes in finite increments; and (2) statistical fluctuations and dependent events force managers to unbalance their plants.

The first of these facts is a generally recognized problem that managers spend endless hours trying to overcome. For example, if a particular process requires 2.58 operators, then it may be necessary to employ 3 operators. If so, then the task at hand typically becomes one of deciding what to do with the 0.42 excess operator capacity. Perhaps there is another work center that can use this excess capacity. As an alternative to having excess slack capacity in the process, overtime might be used instead of hiring the third operator. All of these efforts are directed at trying to prevent paying for excess capacity.

The second fact is not often acknowledged but is no less true than the first. Our previous discussion identifed the potential consequences of statistical fluctuations and dependent events on work-in-process inventories, manufacturing lead times, operating expense, and throughput. Without an appropriate amount of reserve capacity, the throughput, inventory, operating expense, and lead time will be adversely affected, and the firm will be unable to operate efficiently. Ultimately, managers have no alternative but to deliberately run unbalanced plants.

This discussion can be summarized by the following principle of synchronous manufacturing: [1, p. 138]

> *Synchronous Manufacturing Principle 1: Do not focus on balancing capacities, focus on synchronizing the flow.*

BOTTLENECK AND NON-BOTTLENECK RESOURCES

Since a manufacturing plant cannot be balanced, then it follows that some resources will have more available capacity than others. From a capacity standpoint, there are two vastly different types of resources. Resources with excess capacity will be referred to as *non-bottleneck resources*, while those with no excess capacity are termed *bottleneck resources*.

Differentiating Bottleneck
and Non-Bottleneck Resources

The following are more precise definitions of the two different types of resources: [1, pp. 137–138]

Bottleneck Resource—Any resource whose capacity is equal to or less than the demand placed upon it.

Non-Bottleneck Resource—Any resource whose capacity is greater than the demand placed upon it.

To better understand the difference between the two types of resources, consider the case in which a process requires two resources, R1 and R2. Both resources are supposedly manned for 40 hours a week. This represents their potential available capacity. If due to lunch breaks, other personal breaks given to operators, mechanical problems, etc., the resources are normally nonproductive for 1 hour a day, then the available capacity is not 40 hours. The available capacity is actually only 35 hours. Now assume that to produce the mix of products required by the marketplace, R1 and R2 must be productive for 30 and 35 hours per week, respectively. In this example, resource R1 is a non-bottleneck and resource R2 is a bottleneck. (Following the convention first introduced by Dr. Goldratt and used throughout this text, bottleneck resources will be denoted by the symbol X, and non-bottleneck resources will be represented by the symbol Y.)

Recognizing the existence of bottleneck and non-bottleneck resources leads naturally to a development of how time is, or should be, spent at these two different categories of resources. The available time at any resource can be spent in several different ways. [1, pp. 227–228] These may be identified and defined as follows:

Production Time—Time spent processing a product.

Setup Time—Time spent preparing to process a product

Idle Time—Time not used for setup or processing.

Waste Time—Time spent processing materials that cannot be converted into throughput. This may include products of unacceptable quality, work-in-process materials that are not needed, or end items for which there is no demand.

At a bottleneck (X) resource, by definition, all available time should be utilized in production and setup only. Any idle or waste time that occurs at an X resource directly impacts the entire operation. The most serious consequence is the unavoidable loss of throughput. However, at non-bottleneck (Y) resources, production and setup should not consume all of the available time. By definition, after production and setup requirements are accounted for at Y resources, there is still available time left over because there is excess capacity. Hopefully, this excess capacity appears as nothing more damaging than idle time. Some interesting alternatives for effectively utilizing idle time are identified in the next section. But it is most critical to the synchronous operation of the plant that the excess capacity of the Y resources not be transformed into waste time.

It is reemphasized that all manufacturing plants are characterized by unbalanced resource capacities. Therefore, any manufacturing plant might include both X and Y resources. In spite of the general existence of both the X and Y type resources in manufacturing firms, the standard cost approach does not distinguish between these two categories of resources. But X and Y resources are quite different in both their value and significance to the total operation. Understanding these differences is crucial to achieving a synchronized manufacturing environment.

The Value of Resources

When the time components of both bottleneck and non-bottleneck resources are carefully considered, it becomes clear that the value of time at an X resource is fundamentally different from the value of time at a Y resource. The issue of how time is spent at each of these different resources, and the value of that time, is now considered in more detail.

Time Allocation and Value at Bottleneck Resources   Keep in mind that bottleneck resources do not have any excess capacity to spare. By definition, the market-generated demand for productive time at an X resource is equal to or greater than the capacity at that resource. Thus, every second at an X resource must be allocated to either production time or setup time. Any idle or waste time inevitably results in a loss of production time at the X resource. And since the plant can only produce goods as fast as the slowest resources can process them, lost production time at X resources automatically results in lost throughput for the entire plant. Conversely, actions that result in additional productive time at X resources have great value for the entire plant. Additional processing capacity at the critical X resources allows the plant to generate additional throughput.

Suppose, for example, that the amount of setup time at an X resource can be reduced. This may be accomplished through either setup reduction programs that reduce the setup time required for each product or through better scheduling, which reduces the total number of setups. This reduction in setup time requirements frees up time that can be converted into additional production time. As a result, more products demanded by the market can now be processed through the bottleneck with the additional production time that has been created. If there are no other constraints that limit the overall productive capability of the operation, the additional productive capacity at X means additional throughput for the plant.

The same type of reasoning can be applied to any improvement in processing time or other improvements that enable an X resource to process more product in the same available time. Any such improvement at a bottleneck can improve the throughput of the plant. In that case, the value of the improvement is equivalent to the value of the additional throughput that is generated.

Now consider the impact of lost time (due to equipment problems, absenteeism, quality problems, etc.) at a bottleneck resource. At an X resource, every second of available time is needed to try to satisfy the market demand. Any lost time at an X resource will result in both a loss of production time at that resource and throughput for the entire plant. The value of the time lost at the bottleneck resource is equal to the value of the product that could have otherwise been produced.

Time Allocation and Value at Non-Bottleneck Resources   Non-bottleneck resources have more capacity than needed to process the quantity of products demanded by the market. Thus, Y resources are generally characterized by the existence of idle time. This available idle time is the key to determining the significance and value of the Y resources to the production process.

Consider the effect of productivity improvement programs at non-bottleneck resources. For example, what is the impact of a setup reduction program that results in a decrease in the amount of time required to set up a Y resource? The saved setup time cannot be converted to useful production time since we already have more than sufficient time for production. Thus, the saved setup time will only go to increase the idle time available. There is clearly no gain in throughput for the system. Whether or not the setup reduction efforts have any positive effect on the organization at all is questionable. It depends on whether the saved and accumulated idle time is equivalent to at least one increment of capacity. If so, the capacity might be reduced by one incremental unit, and the associated operating expense can be saved. Of course, one incremental unit of capacity usually implies one worker. Thus, in order to realize any direct savings from the setup reductions, the company must be willing and able to fire or lay off workers. Otherwise, the direct cost savings are zero. (It is recognized that setup reductions may result in some indirect benefits for the plant, such as the ability to increase the number of setups and, thus, reduce batch sizes. However, such benefits are not automatic and are not the usual grounds for cost justification.)

Any productivity improvement program at a non-bottleneck resource will generally have little or no impact on the operational measures or the financial bottom line of the company. They do not improve throughput and *may* have only a small impact on operating expense. Therefore, to the extent that improvement programs targeted at Y resources consume managerial or labor time or require investment capital, they are actually counterproductive.

Since non-bottleneck resources have a certain amount of idle time available, any lost time at a Y resource will only impact throughput if the lost time exceeds the available idle time. Even then, lost time at a non-bottleneck resource can often be recovered within a short period of time without adversely affecting the throughput of the system. Thus, the impact of lost time at a non-bottleneck resource is negligible compared to the value of lost time at a bottleneck resource.

The existence of idle time at Y resources, however, can increase the flexibility of the firm. And this flexibility may be used to help improve the process. It is possible to utilize effectively at least some of the idle time at Y resources. One possible use of idle time is to increase the number of setups at the resource. This requires additional setup time, but the resource has excess time to spare. The effect of this action is to enable the resource to process smaller batches. This may have a very favorable impact on reducing work-in-process inventory and on the synchronization of the process. There are many other potential uses of resource idle time. Some of the more typical alternatives are to use the time for worker training or retraining, for conducting maintenance, or for problem identification and problem-solving sessions.

These potential uses of idle time notwithstanding, the excess capacity and resulting catch-up ability of the Y resources play a very significant role in manufacturing processes. It is the excess capacity of the Y resources that gives the manufacturing process the flexibility to function as a synchronous system.

The previous discussion can be summarized by the following principles of synchronous manufacturing: [1, p. 157]

> **Synchronous Manufacturing Principle 2:** *The marginal value of time at a bottleneck resource is equal to the throughput rate of the products processed by the bottleneck.*
>
> **Synchronous Manufacturing Principle 3:** *The marginal value of time at a non-bottleneck resource is negligible.*

Of course, the excess capacity at non-bottleneck resources can be utilized to improve the performance of the manufacturing operation. Principle 3 simply summarizes the fact that non-bottleneck resources already have sufficient excess capacity available to support improved performance. Therefore, increasing the available excess capacity has no significant value.

The realities of the manufacturing environment presented here are in strong contrast to the assumptions inherent in the traditional techniques used to calculate cost savings from productivity improvement programs. Traditional cost techniques assume that reductions in setup time requirements or improvements in processing efficiency translate directly into dollar savings. The dollar amount saved is assumed to be proportional to the dollar cost of providing time at the resource. Furthermore, the cost savings are calculated similarly for both non-bottleneck and bottleneck resources! It should now be evident that the traditional cost approach is incorrect. In the next section, it is demonstrated how to determine the marginal value of time at both X and Y resources.

An Illustration   The ultimate value of any resource has little to do with the direct costs associated with that resource. Indeed, as implied in principles 2 and 3, the true value of a resource is a function of its effect on the throughput, inventory, and operating expense of the plant.

It should be recognized that all actions concerning resources are made at the margin. That is, resource management decisions are designed to either increase or decrease the amount of productive time available at a resource. Therefore, from a managerial perspective, the critical issue is the marginal value to the organization of having more or less capacity at a resource.

The simple production process developed earlier in this chapter and illustrated in Figure 3.1 provides the basis for an example demonstrating the value of resources. Suppose that in this process there are exactly six resources used to produce one product. Forty hours of processing time are available each week at each resource. Resource R3 requires an average of 12 minutes to process each unit of product. Resources R1 and R2 each require an average of 8 minutes to process each unit of product. Resources R4, R5, and R6 each require an average of 6 minutes to process each unit of product. Further assume that market demand for this product varies between 250 and 280 units per week. This situation is presented in Table 3.4.

TABLE 3.4   RESOURCE CAPACITIES FOR THE SIMPLE STRAIGHT-LINE PROCESS OF FIGURE 3.1

RESOURCE	WORK TIME AVAILABLE PER WEEK (HOURS)	PROCESSING TIME USED PER UNIT (MINUTES)	PROCESSING CAPACITY PER WEEK (UNITS OF PRODUCT)
R1	40	8	300
R2	40	8	300
R3	40	12	200
R4	40	6	400
R5	40	6	400
R6	40	6	400

From this information, it appears that 200 units of product can be produced each week. This calculation comes from the fact that resource R3 can process only 5 units per hour for 40 hours each week, yielding a total of 200 units per week. By similar logic, R1 and R2 have the capacity to process 300 units per week, while R4, R5, and R6 have the capacity to produce 400 units per week. Thus, R3 is a bottleneck resource, while the other five resources (R1, R2, R4, R5, and R6) all have excess capacity and are therefore non-bottlenecks.

The relevant financial data for the operation are presented here:

$$\begin{aligned}
\text{Product selling price} &= \$100 \\
\text{Average product cost} &= \$90 \\
\text{Material cost per unit} &= \$20 \\
\text{Direct labor cost/hr} &= \$10
\end{aligned}$$

The marginal value of the X and Y resources can now be determined. To keep the analysis simple, two legitimate assumptions are made. First, the analysis only considers the value of 1 hour of productive time, either added or subtracted, at the X and Y resources. Second, cost savings or expenses incurred as a result of gaining or eliminating productive capacity are ignored. Although these costs can be easily included into the analysis, their effect is negligible and only detracts from the main point of the exercise.

Consider the effect of gaining or losing an hour of time at any of the Y resources (R1, R2, R4, R5, and R6). Gaining an hour of productive time at any or all of these resources only adds to the excess capacity of these resources. There is no gain in throughput since the operation is still limited to producing 200 units per week as determined by the speed of the bottleneck resource. Losing an hour of productive time at any of these resources means only that they will have less excess capacity. Their ability to process the 200 units per week is unaffected by the reduction in capacity. Therefore, since we can add or subtract an hour of capacity from any of the Y resources without affecting throughput, the value of a marginal hour of time at any of these Y resources must be zero.

The impact on the organization of gaining or losing time at the X resource (R3) is quite significant. If an extra hour of productive capacity is realized at R3, then an additional 5 units can be processed by R3. Since the other resources in the operation have excess capacity, this means that throughput will also be increased by 5 units. But what is the ultimate financial impact on the plant? First of all, revenue increases by the total selling price of $500. The only additional costs incurred are the material costs (remember the assumption that no extra labor costs are incurred). At $20 per unit, the total extra material cost is $100. The net result of the additional hour of capacity at R3 is that the plant realizes an increase in net profit of $400. On the other hand, if an hour of productive capacity is lost at R3, then throughput will drop by 5 units, and net profit will decrease by $400. Therefore, it is clear that the marginal value of the R3 resource is $400 per hour.

A key point to remember is that all actions in the manufacturing environment must be considered from the point of view of how these actions affect the throughput, inventory, and operating expense of the entire system. Despite the fact that the standard cost approach does not differentiate between bottlenecks and non-bottlenecks, they obviously are a key consideration in evaluating the ultimate impact of managerial actions. But all the implications of X and Y resources with respect to the synchronous operation of a manufacturing plant have not yet been identified. Thus far we have only seen the tip of the iceberg. The next step in this process is to examine the basic interactions that exist between X and Y resources.

THE BASIC BUILDING BLOCKS
OF MANUFACTURING

The simplest manufacturing environment is one where each resource only works on one product or part. The five basic interactions between bottleneck and non-bottleneck resources that are possible in this type of environment are now identified and analyzed. A critical part of this analysis will focus on the consequences of managing these resources under the traditional manufacturing guidelines of efficiency and utilization as defined in Chapter 2.

The five basic interactions to be analyzed are identified below and labeled as cases 1 through 5:

Case 1: Product flows from bottleneck to non-bottleneck
(Resource X feeds resource Y).

Case 2: Product flows from non-bottleneck to bottleneck
(Resource Y feeds resource X).

Case 3: Product flows from non-bottleneck to non-bottleneck
(Resource Y1 feeds resource Y2).

Case 4: Product flows from bottleneck to bottleneck
(Resource X1 feeds resource X2)

Case 5: Products produced at the non-bottleneck and the bottleneck are both required for assembly
(Resources X and Y feed assembly).

It should be noted that case 4, where one bottleneck feeds another bottleneck, is the least likely of these five scenarios. There are two good reasons for this. First, most operations have a very limited number of bottleneck resources. Therefore, this category of resource interaction is highly unlikely. Second, if such a resource interaction did exist at some time, the resulting consequences are often sufficiently severe to warrant that one of the bottlenecks be broken and changed into a non-bottleneck. Each of the five cases is illustrated in Figure 3.7, and the cases are now examined in turn.

Resource X Feeds Resource Y

In the first case, as illustrated in Figure 3.7(a), material is first processed by a bottleneck and then by a non-bottleneck resource. The X resource must, by definition, work all of the available time to try to meet market demand. Therefore, it makes sense to continue production at X as long as material is available. Managing X to maximum efficiency and utilization, in this case, does not create any problems. But the Y resource is dependent upon the X resource for its supply of material. And Y has the ability to process this material at a faster rate than the X resource. Thus, whatever material is made available by X is processed more rapidly by Y. This causes the Y resource to be starved for work, and the efficiency and utilization measures on Y will appear unsatisfactory. The attempt to manage these resources by the traditional efficiency or utilization criteria presents two major problems. One problem is that the managers of the Y resource are faced with a performance

FIGURE 3.7   THE FIVE BASIC RESOURCE INTERACTIONS

(a) Resource X Feeds Resource Y

(b) Resource Y Feeds Resource X

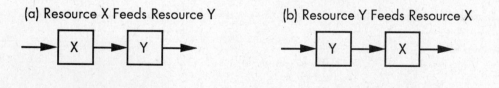

(c) Resource Y1 Feeds Resource Y2

(d) Resource X1 Feeds Resource X2

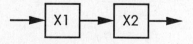

(e) Resources X and Y Feed Assembly

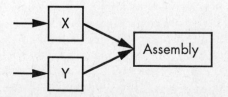

measure with standards by which they cannot excel. A second problem is that the managers' attention may be focused on moving the location of the bottleneck simply to improve the performance of the Y resource.

There is an inevitable contradiction between what is traditionally expected of the managers of the Y resource and what they are really capable of delivering. This contradiction stems from the fact that the traditional methods of evaluating manufacturing systems ignore the effects of resource interactions.

If there were no product and resource interactions with their accompanying consequences, then the manager of each resource could effectively control its level of utilization. However, this will not generally occur since manufacturing is characterized by numerous product and resource interactions. In fact, this discussion illustrates that the effect of a simple interaction between just two resources is far from trivial. An important concept demonstrated by case 1 is summarized by the following principle of synchronous manufacturing: [1, p. 207]

> *Synchronous Manufacturing Principle 4: The level of utilization of a non-bottleneck resource is controlled by other constraints within the system.*

All policies for managing Y resources, as well as criteria for evaluating the managers of these resources, should be consistent with this basic principle. For example, managers of Y resources should not be encouraged to make decisions that are designed to increase the traditional efficiency or utilization performance measures at a non-bottleneck resource. Such actions are often counterproductive.

Resource Y Feeds Resource X

In the second case, as illustrated in Figure 3.7(b), the material is first processed by a non-bottleneck and then by a bottleneck resource. What happens when the resources are managed by the traditional measures of efficiency and utilization? As long as material is available, the Y resource can produce to its full capacity. And under traditional measurement systems, this type of managerial behavior is encouraged. The X resource has less capacity than the Y resource and will take a longer time to process this material. Thus, even though resource X depends on resource Y for work, X will have sufficient material available to maintain good performance as measured by efficiency and utilization. In this case, both resources can maintain good performance against the traditional performance measures.

Since resource Y feeds X, it would appear that principle 4 does not apply. But that is not the case. The issue is really whether or not Y should be restricted to produce at the same rate as X. This issue is related to the question of whether good local performances at X and Y individually lead to good global performance for the entire plant. To evaluate the global performance of the plant properly, the operational measures of throughput, inventory, and operating expense are used. Since the bottleneck is working to its full capacity, the throughput of the system is at a maximum. However, by operating the two resources X and Y to their independent capabilities, excess work-in-process inventory is created in front of X. This is a simple consequence of the fact that the Y resource can produce faster than X. But none of the excess production at Y can be converted into throughput until the material is also processed by X. The operating expense is increased because of the carrying cost of the extra work-in-process inventory. (As will be demonstrated later, the extra inventory also causes lead times to increase and quality to be adversely affected.) As a result, good local performance at Y, as measured by the traditional measures of efficiency and utilization, weakens the overall performance of the entire operation. To improve the performance of the operation as a whole, the production at Y should be curtailed to match production at X.

The term *utilization* conveys a sense of positive contribution. However, production at Y does not always contribute positively to the overall performance of the plant. To distinguish between ordinary production and production that makes a positive contribution to the organization (throughput), we now define the terms *activation* and *utilization*:

> *Activation*—*Refers to the employment of a resource or work center to process materials or products.*
>
> *Utilization*—*Refers to activation of a resource that contributes positively to the performance of a company (throughput).*

Utilization is, by definition, a subset of activation. To be utilized, a resource must be activated, but a resource can be activated without being utilized. When a resource is activated beyond the requirements of utilization, it is said to be overactivated. In traditional usage, the term *utilization* is used to mean activation as defined here; there is no term that means utilization as we have defined it. From this point on in this book, the terms activation and utilization are mostly used as we have defined them. The only exception is when a reference is made to the traditional measure of utilization.

In case 2, the Y resource is being utilized only when its production rate is matched with that of the slower resource X. Any additional production at Y means that resource Y is being overactivated.

The critical concept illustrated by case 2 is summarized by the following principle of synchronous manufacturing: [1, p. 208]

> *Synchronous Manufacturing Principle 5: Resources must be utilized, not simply activated.*

An important point to note is that traditional methods used to evaluate shop personnel typically focus attention on measures that consider only activation instead of utilization. And whenever the process involves material flows from Y to X resources, these performance standards encourage behaviors that result in overactivation, excess work-in-process inventory, and increased operating expense.

Resource Y1 Feeds Resource Y2

In the third case, as illustrated by Figure 3.7(c), the material is first processed by the non-bottleneck resource Y1 and then processed by the non-bottleneck resource Y2. Since Y1 and Y2 are both non-bottlenecks, both can produce at a faster rate than required to support market demand. When resources such as Y1 and Y2 are managed by the traditional measures of efficiency and utilization, the result is likely to be overactivation. In fact, traditional performance measures encourage such overactivation.

One clear result of the overactivation of Y1 and Y2 is the creation of excess inventory somewhere in the system. Depending on the relative productive capacities of Y1 and Y2, the major accumulation of inventory may be either between the two resources or beyond Y2. However, regardless of the relative capacities, some excess inventory will always be found beyond Y2. This inventory may exist as primarily work in process or finished goods, but it will exist.

The actual output of resources Y1 and Y2 should be matched to market demand or to other limiting constraints in the system. Encouraging overactivation, as traditional measurement systems usually do, only increases excess inventory. This does not contribute to the throughput of the plant, but instead detracts from the performance of the entire operation. Thus, synchronous manufacturing principles 4 and 5 are valid even under the conditions of case 3, where a non-bottleneck feeds a non-bottleneck.

Resource X1 Feeds Resource X2

In the fourth case, as illustrated by Figure 3.7(d), the material is first processed by the bottleneck resource X1 and then by the bottleneck resource X2. Since both resources X1 and X2 are bottlenecks, neither can produce at a rate fast enough to keep pace with market demand. As a result, one might mistakenly conclude that neither resource can be overactivated and that the traditional measures of efficiency and utilization would not cause any harm in this case. However, remember that resources do not work in isolation. Resources must be evaluated based on how they impact the overall flow of material in the system. There are only two basic scenarios to consider.

Consider the case for a product where resource X1 produces at a slower rate than resource X2. Resource X2 is continually starved for material and can only produce as fast as resource X1. Because the production rate at X2 is less than market demand, there should be no buildup of inventory in the system beyond X2. Furthermore, since resource X2 cannot be fully utilized, X2 actually has excess capacity. Thus, even though resource X2 is a bottleneck by definition, it does not limit the ability of the system to produce goods. In fact, in this case, resource X2 takes on the characteristics of a non-bottleneck resource. For this scenario, the interactions between resources and the impact on the system is similar to those described in case 1 (resource X feeds resource Y).

Next consider the case for a product where resource X1 produces at a faster rate than resource X2. In this case, the total throughput of the system is clearly controlled by resource X2. Even though resource X1 is a bottleneck (capacity less than demand), X1 takes on the characteristics of a non-bottleneck resource as described in case 2. Given the interactions between resources, X1 can be fully activated at all times but cannot be utilized all the time. The processing capabilities of resource X2 effectively limit the utilization of resource X1.

These two scenarios again demonstrate that interactions between resources and products significantly affect the proper management of the resources. Even the management of bottleneck resources must be based upon the degree to which the bottlenecks affect the overall flow of materials in the total system. Each bottleneck resource must be managed differently, according to its impact on the system.

Resources X and Y Feed Assembly

In the fifth case, as illustrated in Figure 3.7(e), the emphasis is not on the flow of material between X and Y resources. Instead, the focus is on the situation where X and Y resources process different materials that are then combined together at an assembly operation. Resource X must produce continuously in an attempt to keep pace with market demand and, therefore, cannot be overactivated. But resource Y is capable of producing at a faster rate than that required by market demand, and certainly at a faster rate than can be supported by resource X. Therefore, resource Y can be overactivated. Once again, it is observed that managing to satisfy traditional performance measures is likely to encourage such overactivation of Y.

The assembly operation can proceed no faster than the rate at which both component parts are made available. Thus, assembly is constrained to produce at the rate of the slower X resource. If Y is overactivated, some of the material processed by Y will accumulate in front of the assembly operation as component-part inventory. In fact, any production at Y in excess of the production at X only creates excessive work-in-process inventory of Y parts in front of the assembly operation.

It should be clear that interactions between resources may exist in a manufacturing environment even without a direct connection between the resources. In the terminology of synchronous manufacturing, two resources interact with each other if the activity at one resource influences either the activity or the result of the activity of the other resource. In case 5, even though there is no direct flow of material between X and Y, it is evident that there is a significant interaction between the resources. In fact, the logical conclusion is that both throughput and the utilization potential of resource Y are again controlled by resource X.

The analysis of the five basic ways in which resources interact has led to the identification of five important principles of synchronous manufacturing. Two additional principles will be identified in Chapter 5. These principles provide a strong foundation for the development of the synchronous manufacturing philosophy.

SUMMARY

Two basic phenomena—dependent events and statistical fluctuations—are common to all manufacturing operations. The simultaneous presence of both of these phenomena in a manufacturing environment has a very serious consequence. Because of the existence of dependent events, variances in the product flow caused by statistical fluctuations do not average out. Instead, the negative variances accumulate, disrupting the planned product flow for the entire plant.

Given the inherent nature of resources, it is virtually impossible to balance the capacity of resources in a manufacturing plant. But this is actually a blessing in disguise. Because of the numerous disruptions that consistently plague all manufacturing plants, attempts to balance the capacity of a plant are often counterproductive. Such balancing activities may adversely affect throughput, inventory, and operating expense for the entire plant.

Because resource capacities cannot be balanced, two different categories of resources can be identified—bottlenecks and non-bottlenecks. All of the capacity at bottleneck resources is required for processing throughput. There can be no lost time at a bottleneck or throughput suffers. But non-bottleneck resources have excess capacity. As a result, the marginal value of processing capacity at a bottleneck is extremely valuable, but the marginal value of processing capacity at a non-bottleneck is essentially zero up to the point where all of its excess capacity may be consumed. Whether a resource is a bottleneck or a non-bottleneck makes a significant difference in the way the resources should be managed.

The interactions that exist between resources lead to the conclusion that there is a distinct difference between activation and utilization. In many cases, it is possible to activate a resource, especially a non-bottleneck, beyond what is useful or productive for the system. But according to the synchronous manufacturing philosophy, resources must only be utilized (i.e., activated to contribute positively to company performance), not simply activated. The level of utilization that is possible for a non-bottleneck resource is limited by the system. And activating a resource without utilizing that resource to help achieve plant objectives is both wasteful and costly.

QUESTIONS

1. Define dependent events and statistical fluctuations. Give examples of each in a manufacturing environment.
2. What are the effects of dependent events and statistical fluctuations on throughput, inventory, and operating expense?
3. Give additional examples of systems that are subject to dependent events and fluctuations. Do they exhibit the phenomenon of accumulating deviations? Explain.
4. Discuss synchronous manufacturing principle 1 as applied to a column of marching soldiers.
5. Why is it necessary for management to know which resources in the plant are bottlenecks and which are non-bottlenecks?
6. Explain the difference between activation and utilization.
7. Describe activation and utilization as they relate to the following cases:
   a. A raw material is sequentially processed through resources Y1, X, and Y2 to become a finished good.
   b. Raw material 1 is processed at resource Y1 and raw material 2 is processed at resource Y2. The two partially processed materials are assembled at a non-bottleneck resource to become a finished good.
8. Describe how you might manage if the supply source is the bottleneck.
9. Describe how you might manage if the demand source is the bottleneck.

PROBLEMS

1. Construct an example in which two resources in sequence in a manufacturing process show random fluctuations from a production schedule. Demonstrate the accumulation of deviations.
2. Product P1 is produced by processing material through four resources (R1, R2, R3, R4) in sequence. There are 480 minutes of processing time available per day at each resource. Product P1 requires 20, 24, 30, and 16 minutes of processing time per unit at resources R1, R2, R3, and R4, respectively. Demand for P1 is 20 units per day. Categorize each of the resources as either a bottleneck or non-bottleneck.
3. Refer to problem 2. If the selling price for P1 is $100 per unit and the material content of P1 is $40.00, calculate the marginal value of 1 hour at each of the four resources. (Hint: Consider resource R2 carefully.)

4. For product P1 in problem 2, derive a typical 5 day schedule for resources R1, R2, R3, and R4 that is based on the traditional focus of efficiency. Compute the resulting efficiency (activation) of each resource. Also discuss the resulting levels of inventory that would result using this approach.

5. For product P1 in problem 2, derive a typical 5 day schedule for resources R1, R2, R3, and R4 based on a focus of utilization. Compute the resulting efficiency of each resource. Discuss the resulting levels of inventory that would result using this approach. Compute the plant (system) utilization for the week.

4

Identifying and Managing Constraints

THE SIGNIFICANCE OF CONSTRAINTS

There is a growing awareness among manufacturing managers of the necessity of achieving a smooth and fast flow of materials through the operation in concert with market demand. But few organizations have made significant progress toward that goal. The primary reason for the lack of progress is neither a lack of concern nor a lack of effort. In fact, almost all organizations have expended large amounts of time and money implementing a variety of programs designed to improve the productivity of the operation. Nevertheless, most managers find themselves disappointed with the results. The lack of progress can be traced to two main causes: [3, p. 1]

1. Most managers do not have a clear understanding of how to achieve a synchronized product flow in a complex and dynamic manufacturing environment.
2. There is a basic conflict between the requirements for a synchronized flow and the existing infrastructure of management practices and policies.

A primary contributing factor to these two problems is management's inability to identify and properly control the constraints that exist in every organization. In order to fully develop a logical approach to synchronizing manufacturing operations, it is first necessary to explain the significant role played by constraints in a manufacturing environment.

Managers of manufacturing plants would be the first to admit the existence of constraints in organizations. However, even the most astute manager may not fully realize the degree to which various constraints affect the operation of the organization. To provide a common point of reference, the concept of a constraint is defined in terms of the goal of the organization:

> A *constraint* is any element that prevents the system from achieving the goal of making more money.

Every organization has at least one constraint. Otherwise the firm would be able to make an unlimited amount of money. The degree to which any system can perform is ultimately determined by the set of constraints that govern the system. Therefore, to improve the productivity and profitability of the firm, managers must focus on the constraints that limit the performance of the system. Fortunately, this is not a battle that must be fought on a large number of fronts. In most firms, there are a very small number of constraints that limit current performance. However, in a dynamic manufacturing system, especially one where improvements are being made, the constraints are likely to change over time.

It is management's job to try to limit the adverse effects that constraints have on the firm's productivity and profitability. But before a significant program of focused improvements can be developed, it is first necessary to identify and understand the various types of constraints that are inherent in manufacturing systems.

TYPES OF CONSTRAINTS

There are several categories of constraints that exist in manufacturing environments. Included are market, material, capacity, logistical, managerial, and behavioral constraints. The requirements and needs of the market define the throughput limits for the firm. Material and capacity problems within a production process are highly visible to production managers. Thus, real or perceived material and capacity constraints are likely to receive a great deal of attention. Logistical, managerial, and behavioral constraints also exist in manufacturing environments but are often unrecognized as constraints to the process. However, these three constraints are often responsible for disruptions in the production process that are mistakenly attributed to material and capacity constraints.

Market Constraints

In any manufacturing operation, market demand is the critical driving factor. The demands of the marketplace determine the throughput boundaries within which the firm should operate. The type of product for which there is demand is determined by the marketplace. Additional considerations such as quantity limits, lead time requirements, competitive pricing, and quality standards are not typically determined by the firm, but rather by the market.

Are market constraints a significant problem? To answer this question, just consider the billions of dollars of "dead" or obsolete inventory stocks that fill the shelves of warehouses across the country. Moreover, any sales manager can probably recall numerous instances of lost sales because of a firm's inability to provide the desired product in the required time.

Management's task is to plan an efficient and manageable production flow that will yield products that meet the market requirements identified above. In a very real sense, the ultimate constraint on a manufacturing firm is the market. If the firm cannot satisfy the demands of the marketplace, the firm will not survive.

Material Constraints

Without the necessary material inputs, the manufacturing process must shut down. This truism has been recognized by managers as long as there have been production processes. In fact, the necessity of having sufficient raw materials and work-in-process materials to keep the production process flowing has given rise to a large variety of material control systems. Many of these systems are designed to guarantee an overabundant supply of material. Unfortunately, such systems usually create more problems than they solve.

Material constraints can be considered to be either short term or long term in nature. Short-term material constraints frequently result when a vendor does not deliver as scheduled or the material is defective. Such situations have great potential for disrupting the smooth flow of the production system. Material constraints can also result from an inadequate planning horizon. When insufficient visibility into the future is coupled with long purchasing lead times, material problems are inevitable. Long-term material constraints are typically the result of a material shortage in the marketplace. In such a situation, the availability of quality material and the lead time required to obtain the material are major concerns. Material constraints that constitute a continuing limitation on the production process should be considered when developing the master schedule.

Material constraints may also develop during the production process as a result of insufficient work-in-process inventory components. This problem may usually be traced to one of four causes. Material shortages are often the result of poor scheduling of the product flow. For example, a work station may alternately be overloaded, then starved for material. Material shortages may result because a particular operation in the production process has produced an excessive amount of scrap or defective units that cannot be reworked. As a result, there is insufficient material to complete the order at succeeding operations. Material shortages may also occur when a feeder station is down and unable to supply necessary material to downstream stations. Finally, material shortages may also be the result of "stealing." Stealing occurs when material that has been designated for a particular order or product is diverted to a different product. Regardless of the reasons for the reallocation of the material, a material shortage for one or more products is automatically created. The problem of material misallocation will be discussed in detail in a later chapter.

Capacity Constraints

There are two key factors that directly influence a plant's ability to maintain the required production flow in a smooth and timely manner. One factor is the availability of material, already discussed. The other factor is the availability of capacity. A capacity constraint is said to exist when the available capacity at a resource may be insufficient to meet the workload necessary to support the desired throughput. The result is a potential disruption of the product flow.

When managers are asked to identify constraints within their production process, capacity constraints are usually the first to come to mind. That is, there are certain resources which, because of an apparent lack of available capacity, cause disruptions to the smooth flow of products through the plant. These resources, when they can be identified, tend to receive special attention from managers.

The distinction has already been made between bottleneck and non-bottleneck resources. Recall that a non-bottleneck resource was defined as any resource whose capacity is greater than the demand placed on it. A bottleneck resource was defined as any resource whose capacity is equal to or less than the demand placed on it. If there are bottlenecks in the operation, then the actual flow of products through the plant is going to be less than the desired product flow unless the capacity of the bottleneck resources can be increased. Most experienced managers quickly identify with the term *temporary bottleneck*. That is, at a given point in time, one resource will appear to be the bottleneck, and at a later time, that same resource will appear to have excess capacity. In some plants, this phenomenon appears to move around from one resource to another. Resources almost seem to take turns being the bottleneck. The term *wandering bottlenecks* is used to describe this syndrome.

When a bottleneck exists in an operation, both the throughput and the timely completion of products are in jeopardy. However, even in plants that have no true bottleneck resources, there are usually one or more resources that have the potential to cause a major disruption in the timing of the product flow. These resources are called capacity constraint resources (CCRs) and are discussed in depth in a later section of this chapter.

The problems of lost throughput and missed due dates that continually plague many manufacturing plants may be attributed to the lack of processing capacity of one or more resources. In some plants, this is clearly the case. Often, however, the problems may be directly attributable to the mismanagement and poor scheduling of one or more critical resources. Many of these problems are also caused or aggravated by the existence of other non-capacity constraints.

Logistical Constraints

Any constraint that is inherent in the manufacturing planning and control system used by the firm is referred to as a logistical constraint. The primary effect of this type of constraint is that it acts as a drag on the smooth flow of goods through the system. Logistical constraints may adversely affect the synchronous operation of the system at any point from order entry to final shipment. These constraints are essentially built into the manufacturing system and may be difficult to change. In fact, logistical constraints are not typically recognized by managers as parameters that can be manipulated. However, if the constraints imposed by the planning and control system are too severe and significantly disrupt the synchronous flow of products, then the system must be modified or changed.

To illustrate, consider an order entry system that utilizes order taking at the local level by sales representatives. Orders are collected and eventually forwarded to corporate headquarters where they are combined with other orders from across the country. Finally, the orders are processed and master production schedules are developed for each of several plants in various parts of the country. Such a process may require several weeks simply to secure orders and parlay that information into production schedules. The eventual receipt of the order by the customer will be delayed by the length of the order-processing lead time.

Another illustration of logistical constraints is evident in material control systems that use monthly time buckets. By using time buckets of 1 month's duration instead of 1 week or 1 day, there is a loss of visibility of the exact due dates for orders. This loss of visibility means that if all orders are to meet their promised due dates, then some orders may be completed up to 4 weeks early. In addition, the total lead time will be excessive. The total amount by which the customer's lead time is needlessly increased depends on the actual manufacturing system.

These two situations illustrate the effect of logistical constraints in manufacturing systems. The first case involves an inefficient order entry system. The second case involves a material control system with lengthy time buckets. In both cases, the resulting lead time that can be promised to customers is much greater than would otherwise be required. This clearly has an adverse effect on the ability of the firm to run a synchronous operation.

Managerial Constraints

Managerial constraints are the management strategies and policies that adversely affect all manufacturing-related decisions. In many cases, managerial constraints are the result of a lack of understanding of the factors that enhance or detract from a synchronous operation. Managerial constraints may impact the system in two basic ways. They may create situations that lead to a suboptimization of the system, or they may compound the effect of other constraints that exist in the system. These two situations are illustrated here.

Managerial constraints may have the effect of magnifying the problems caused by the other constraints in the system. One example of this is the policy of determining batch sizes by using an economic order quantity. It has already been established that the EOQ approach is not appropriate for making batch-size decisions in a manufacturing environment. Another example is the practice of letting supervisors independently sequence jobs at non-bottlenecks in order to save time on setups. Given the existence of bottlenecks or CCRs in the system, the poor schedules typically generated by either of these policies will likely disrupt the smooth and timely flow of products through the system. As a result, throughput could be lost. Furthermore, production lead times are likely to be excessive, jeopardizing shipping due dates.

Managerial constraints may also be responsible for creating situations that encourage behaviors that lead to global suboptimization. For example, consider a firm that gives bonuses to purchasing agents based on positive purchase price variances. Normally, the buyer would purchase sufficient steel to satisfy the needs of the process for 1 month at a time. But instead, in order to receive a price break, the buyer purchases a 6 month supply of steel. The price of the steel (6 months' worth) is $3 million. This includes a $50,000 price break. In the period of 1 year, based on the quantity discounts received, it appears that the buyer would save the firm $100,000 (less performance bonuses paid to the buyer). But now consider what the firm must pay to carry the extra inventory. If 6 months' worth of steel is purchased at a time, the value of on-hand inventory averages $1.5 million during the year. If 1 month's worth of steel is purchased at a time, the value of on-hand inventory averages only $250,000. If the firm has an inventory carrying cost of 30 percent, then the extra inventory will cost the firm $375,000 per year [.30 × ($1,500,000 − $250,000) = $375,000]. Thus, the firm saves $100,000 per year in purchase price but spends an additional $375,000 to carry the extra inventory. The buyer, acting in accordance with the management policy that encourages positive purchase price variances, costs the firm $275,000 a year.

Behavioral Constraints

To a certain degree, organizations can be characterized by the attitudes and behaviors exhibited by the workforce. To the extent that behaviors are developed and practiced that run counter to the principles of synchronous manufacturing, these behaviors become a constraint on the system. Behavioral constraints may be generated from the work habits, practices, and attitudes of the managers or workers. These attitudes often reflect the mores and culture of the entire organization. In most manufacturing environments, behavior patterns evolve as a result of the management style practiced in conjunction with the performance evaluation and reward structure that backs it up. Thus, management may be at least partially responsible for many of the behavioral constraints found in firms.

One example of a behavioral constraint is the "keep busy" attitude often exhibited by many supervisors and workers. This attitude may be generated by the fear that if managers can't keep their workers busy, they may lose them. This attitude is quickly picked up by the workers. To the extent that a plant has excess capacity, the result of this attitude is that work may be completed that is neither scheduled nor required. This practice does not add to the throughput of the plant; it only generates excess work-in-process and finished goods inventory.

Another example of a behavioral constraint is the tendency on the part of shop operators to cherry-pick from among the jobs that are waiting in the queue of a work station. Cherry-picking involves the practice of choosing the easiest or most desirable jobs to process first, leaving the less desirable jobs until later. This practice may become especially prevalent in plants where there is more than one shift. One of the many problems with this practice is that it tends to destroy the schedule, which results in unpredictable lead times and missed due dates.

Behavioral constraints may not be the primary cause of problems within a plant. However, where they exist, they are often difficult to remove. Thus, behavioral constraints often constitute a major obstacle to improving the process.

Other Considerations

Identifying and properly controlling all of the various constraints in a manufacturing firm is a virtual impossibility. Even as constraints are identified and brought under control, new constraints may develop. The task of constraint management is a never-ending process. Moreover, no matter how well the production flow is planned, the actual flow of materials through the plant will differ from the planned flow. This is the inevitable result of three key factors that affect all manufacturing operations:

1. Unpredictable disruptions, such as machine breakdowns. For example, even if it is known that a particular work center is down an average of 20 hours a month, it is impossible to predict when it will be down.
2. Inaccurate/indeterminate information, such as time standards. The precise time required to set up and/or process a batch generally varies, although the average over several batches may match the standard. Also, in most instances, the process time standards contain inaccuracies or are not complete.
3. Large numbers of variables. The number of variables that can influence the performance of a given work center is so large that many of these variables are not considered in the planning process.

As a result, the actual production flow in microscopic detail—i.e., the exact resource on which the material is processed, the exact time at which it is done, the time it takes to process the batch, the quantity processed, etc.— will not match the planned flow. All of the various constraints that affect the planned product flow will naturally tend to interact in such a way as to disrupt the planned product flow. The net result is likely to be missed due dates and suboptimization of throughput, inventory, and operating expense. Synchronous manufacturing is a management philosophy that can be utilized to control the adverse effects generated by constraints in a manufacturing environment.

CAPACITY CONSTRAINT RESOURCES

CCR Defined

The concept of capacity constraint resources (CCRs) is extremely critical in the synchronous manufacturing philosophy. [5, p. 98] The following discussion provides the necessary groundwork for establishing the proper role of capacity constraints in a synchronous manufacturing environment.
The term *capacity constraint resource* is formally defined as follows:

> **Capacity Constraint Resource**—*Any resource which, if not properly scheduled and managed, is likely to cause the actual flow of product through the plant to deviate from the planned product flow.*

The Relationship between Bottlenecks and CCRs

It is probably not clear from this definition whether a CCR is a bottleneck or a non-bottleneck resource. Depending on the situation, a CCR may be either. In order to further clarify the concept of CCRs and how they relate to bottleneck and non-bottleneck resources, consider the situation illustrated in Table 4.1.

An Illustration   In the simple process described in Table 4.1, products A and B are manufactured by processing them through resources R1, R2, R3, and R4, in that sequence. Demand averages 2 units of product A and 5 units of product B per day. The resource capacity required to support that level of demand is 31 hours of R1, 24 hours of R2, 23 hours of R3, and 7 hours of R4 per day. There are only 24 hours of capacity available per day. Therefore, by definition, resources R1 and R2 are bottlenecks since the processing time required at these resources equals or exceeds the available processing time per day. Conversely, resources R3 and R4 are non-bottlenecks, since their processing time requirements are less than the available capacity.

But which, if any, of the four resources are CCRs? That is, which of the four resources are likely to disrupt the planned flow of product through the plant if they are not properly managed and controlled?

TABLE 4.1   AN ILLUSTRATION TO DEMONSTRATE THE RELATIONSHIP BETWEEN BOTTLENECKS, NON-BOTTLENECKS, CCRs, AND NON-CCRs

| | AVERAGE DEMAND PER 24 HOUR DAY | HOURS OF RESOURCE TIME REQUIRED TO PRODUCE ONE UNIT OF PRODUCT | | | |
		R1	R2	R3	R4
Product A	2 Units	3	2	9	1
Product B	5 Units	5	4	1	1

| | | REQUIRED AND AVAILABLE CAPACITY | | | |
		R1	R2	R3	R4
	Required Capacity per Day (hours)	31	24	23	7
	Available Capacity per Day (hours)	24	24	24	24

Let's first consider the role of bottleneck resources. For a given process configuration, the actual product flow is not controlled by all of the bottlenecks. Some are more severe than others. The actual product flow through the plant is controlled by the most severe bottleneck and not necessarily by the others. In this example, the flow of material to resource R2 is constrained by the fact that resource R1 processes material at a slower rate than R2. As long as the capacity at resource R1 is less than at resource R2, then R2 does not directly control the flow of product through the process. Any product that gets through resource R1 should have no trouble getting through resource R2.

Since resource R1 has the greatest capacity requirement and this requirement exceeds the available capacity, resource R1 is the primary constraint on the product flow. Therefore, any mismanagement at resource R1 that results in lost processing time causes a reduction in throughput and may adversely affect shipping due dates. Even though resource R2 is a bottleneck, it is not an active constraint on the process and need not be considered when planning the product flow. It is sufficient to consider only resource R1. Therefore, resource R1 is a CCR, while resource R2 is not.

Another important consideration is that the product flow involves not only an element of quantity, but an element of time as well. Bottlenecks, of course, control the quantity produced. But there may be resources that are non-bottlenecks, which nevertheless can interrupt the timing of the product flow. When planning the product flow, i.e., scheduling the system, these other resources must be considered. Otherwise, the actual product flow out of the factory will not match the scheduled flow. This can impact the plant's ability to ship products as promised.

To illustrate this situation, consider a simplified example. Suppose the plant receives an order for 20 units of product A and 50 units of product B. The order is due in 11 days (264 hours). To simplify, assume the plant currently has no work-in-process orders requiring the use of resources R1, R2, R3, and R4. Thus, the plant is free to begin immediately processing the latest order. In order to help process the order, suppose that management decides to subcontract for 7 hours of additional processing time on resource R1 per day for the next 10 days. With the subcontracting, do all the resources appear to have sufficient capacity to process the order on time? (Try to determine the answer before moving on.)

Since the order quantity is exactly 10 times the average daily demand, the total capacity required to produce the order is also exactly 10 times the daily requirement. Therefore, in total, resources R1, R2, R3, and R4 require 310, 240, 230, and 70 hours to process the order. Actually, since 70 hours of R1 time is subcontracted, R1 only requires a total of 240 hours. Since the order is due in 264 hours and we can start immediately, it might appear that there is no problem in getting the order out on schedule.

Suppose the order is processed by scheduling all 20 units of product A first, followed by all 50 units of product B. Further suppose that the batches are overlapped so that as soon as the first unit of a batch is processed at a resource, that unit is immediately moved to the next resource for processing. This schedule results in a total completion time of 247 hours. (Resource R1 completes its work in 240 hours, R2 finishes its work 4 hours later, R3 finishes 1 hour later, and R4 finishes 1 hour later, for a total completion time of 246 hours.) The order is completed with 18 hours to spare before the scheduled shipping due date.

But what if the schedule calls for the 50 units of product B to be processed first, followed by the 20 units of product A? With this schedule, the order is doomed to be late. Why? In the first 8 days, resource R1 has 248 hours of available capacity (8 days @ 31 hours per day). The 50 units of product B require 250 total hours. Therefore, the last unit of product B is finished at R1 in 8 days plus 2 hours. This converts to 194 clock hours (8 days @ 24 hours per day + 2 hours). Therefore, processing on product A at R1 can begin after 194 hours. The first unit of product A is completed at R1 after only 3 hours, whereupon it is moved to resource R2. Resource R2 finishes the first unit in 2 hours and sends it to R3. Now, resource R3 requires 9 hours of processing time per unit for 20 units of product A. Resource R3 requires 180 hours of processing time. Thus, R3 is finished with the order after a total of 379 hours have passed: $(194 + 3 + 2 + 180 = 379)$. Finally, resource R4 finishes the last unit 1 hour after R3 is completed. The total processing time has consumed 380 hours. The order is 116 hours late.

What happened? Clearly the order is late because the schedule caused resource R3 to be mostly idle for a long period of time when product B was being processed. When product A finally begins to be processed, a huge backlog of work quickly accumulates in front of R3. By the time R3 can process all the work, the order is way overdue. Clearly, resource R3 meets the requirements of a CCR.

In some cases, non-bottleneck resources play a vital role in the product flow. In the example illustrated in Table 4.1, the most loaded resource is clearly R1. However, when producing product A, the constraint to the product flow is not resource R1, but resource R3. When this fact is ignored and the schedule is based only on the ability of the bottleneck resource to process goods, then poor scheduling or mismanagement of resource R3 may result in a disruption of the planned product flow and missed due dates. To obtain realistic shipping dates, resource R3 must be considered, even though it is not a bottleneck for the entire system.

Most managers can testify that resources which should have sufficient capacity to meet the processing requirements often unexpectedly pop up and act as a constraining resource in their manufacturing process. Such resources are very similar to resource R3 as described in Table 4.1.

Of course, most non-bottleneck resources are not CCRs. For example, resource R4 has plenty of excess capacity and requires less processing time per unit of product than either R1 or R2. Furthermore, it is not a constraint in producing either product A or product B. As a result, resource R4 is not a CCR. This resource need not be considered when planning the product flow.

In summary, the example illustrated in Table 4.1 demonstrates all four possible combinations of bottleneck/non-bottleneck resources with CCR/non-CCR resources. The resource combinations are summarized in Figure 4.1.

FIGURE 4.1   A CLASSIFICATION OF RESOURCES REPRESENTED IN TABLE 4.1

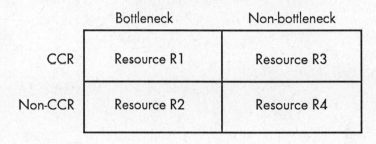

	Bottleneck	Non-bottleneck
CCR	Resource R1	Resource R3
Non-CCR	Resource R2	Resource R4

The Effect of CCRs on the Product Flow   The expected effects of the four basic categories of resources on the product flow are summarized in Figure 4.2. The main point of this illustration is that managers must concentrate their attention on those resources that are CCRs. Whether or not a resource is a bottleneck is not the critical characteristic. For purposes of control and to ensure maximum throughput from the process, the CCRs are the critical resources. Therefore, from now on, our focus will be on CCRs instead of bottlenecks.

It is reemphasized that in most manufacturing firms today, there are few, if any, true bottleneck resources. However, even in plants that have no bottleneck resources, there are usually one or more CCRs. Our experience indicates that in many plants, resources that may appear to be bottlenecks are quite often non-bottleneck resources. But a thorough analysis usually reveals that these troublesome non-bottleneck resources are CCRs.

Identifying the CCRs

The traditional approach to identifying capacity problems in a manufacturing plant involves the use of resource load analysis. The identification and significance of CCRs can be further conceptualized and explained in a format generally used by the load analysis approach. [3, p. 16–17]

FIGURE 4.2 THE EXPECTED EFFECTS OF THE FOUR BASIC RESOURCE CATEGORIES

	Bottleneck	Non-bottleneck
CCR	Will constrain actual flow, both in quantity and time. Must be considered in planning the product flow.	Will constrain the timing of the actual flow, but not the quantity. Must be considered in planning the product flow.
Non-CCR	May constrain actual flow, both in quantity and time. Need not be considered in planning the product flow.	Does not constrain the flow, either in quantity or in timing. Need not be considered in planning the product flow.

Resource Load Profiles It is not difficult to understand how a bottleneck resource can be a CCR. It is our purpose to demonstrate again how a non-bottleneck resource can also be a CCR.

Consider Figure 4.3, which identifies the planned workload of resources R1, R2, and R3 that is required to support the ideal product flow plan. Suppose that the relevant time period for planning purposes in this manufacturing system is 1 week (5 days). Each figure depicts a planned workload for a resource covering two planning periods. The loads are presented in terms of fluctuating daily requirements, which are derived from the particular mix of products scheduled to be processed on a given day. The capacity requirements are also graphically compared to available capacity.

Figure 4.3(a) indicates that the required daily capacity for Resource R1 is always less than the available capacity at that resource. If the information on which this planned capacity is based is reasonably accurate, then it can be concluded that resource R1 does not represent a capacity constraint. Thus, the available capacity of resource R1 is not a factor that must be considered in planning the production flow.

Figure 4.3(b) shows the planned capacity requirements for resource R2. It is clear from the figure that the average required capacity of resource R2 is less than the available capacity. However, the daily workload required to support the planned product flow exceeds the available capacity on days 3, 6, and 8. If the information is accurate and the material actually arrives

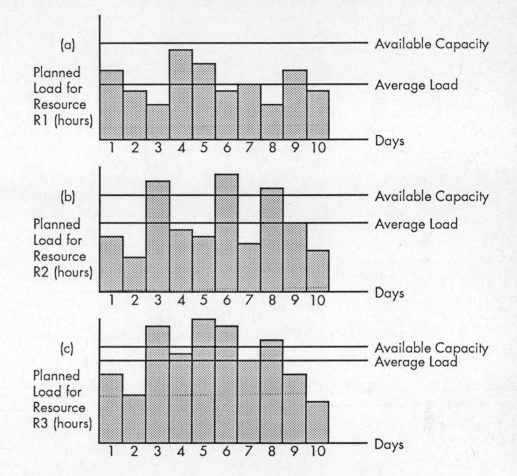

according to plan, then on these days, R2 will not be able to complete the planned work. Some of the work scheduled for day 3 will be carried over to day 4. By examining the planned loads for days 3 and 4, it is clear that there is sufficient capacity to complete all of the planned and backlog work by the end of day 4. The same is true of the overloads that occur on days 6 and 8. In each case, the overload is immediately made up the next day. In essence, the daily fluctuations in the load average out within the weekly planning period. The load fluctuations result only in minor fluctuations of the planned product flow on a daily basis. But significantly, these fluctuations are immediately made up, and the weekly planned product flow is not disrupted. The appropriate conclusion is that the available capacity of resource R2 is not a factor that must be considered when planning the production flow.

Figure 4.3(c) illustrates the planned capacity requirements for resource R3. Once again, the average load required to support the planned product flow is less than the available capacity of the resource. But in this case, there is not a large amount of excess resource capacity to cover for the daily fluctuations in the workload. The capacity required by the plan exceeds the available capacity at R3 on days 3, 5, 6, and 8. Any uncompleted work scheduled for those days will be carried over to the following days. However, unlike the previous case, there is insufficient capacity to process both the planned work and the overflow work on day 4. The work not completed on day 4 will carry over to day 5. However, on day 5, the scheduled work itself exceeds the available capacity. In fact, a close examination of Figure 4.3(c) indicates that the accumulated work increases at least through day 6. Moreover, the work that was planned to be completed in the first week cannot be completed. There will be a disruption of the planned weekly product flow, caused by the limited capacity at R3. Therefore, resource R3 is a capacity constraint resource. The available capacity of R3 must be considered when planning the product flow.

Once the capacity constraints have been identified, the planned product flow must be constructed in a way that recognizes the limitations of these constrained resources. It is not possible to prevent the inevitable minor fluctuations between the actual and the planned product flows for individual resources. However, we cannot allow these minor fluctuations to cause the aggregate product flow to deviate from the plan. The resulting production plan, which considers the CCRs, becomes the basis for scheduling actual production and for making promises to customers.

An Alternative Procedure In principle, a master production schedule can be used to generate the load for a particular resource required to support the desired product flow. This resource load profile can then be compared to the available capacity to determine whether the resource is a bottleneck or a CCR. However, the practical application of this method has severe problems. The most serious of these problems is the integrity of the data available to perform the load analysis. The outcome of this analysis is highly dependent upon critical information, such as the product mix, estimated lead times, setup and processing time requirements, and inventory availability. But in almost all plants, the data describing these types of information contain gross errors. These errors are severe enough that the data cannot be used for load analysis and CCR identification without first undertaking a mammoth data cleanup task.

Fortunately, it is not necessary to get bogged down in a massive data cleanup effort. Load analysis is not absolutely required to identify the CCRs in the plant. The systematic procedures contained in the synchronous manufacturing philosophy can perform the same task quite effectively. The CCRs and bottlenecks can be identified without the use of the data typically generated and used in the load analysis approach.

Instead of using the typical load analysis procedure to identify capacity problems, significant benefits can be derived by reversing the procedure. Instead of using the load analysis data to identify the CCRs, the knowledge of which resources are CCRs can be utilized to identify the areas where it is most important to have accurate data. In keeping with the spirit of the synchronous manufacturing approach, it is only necessary to be actively concerned about data that can seriously impact the total system. The severity of data errors can be best judged by the ultimate impact of inaccurate information on the throughput (T), inventory (I), and operating expense (OE) of the total system. Thus, the data cleanup effort should be concentrated on those resources that have a significant impact on T, I, and OE.

The objective of the resulting data cleanup effort is not to have fully accurate data everywhere in the system. Although this may appear to be a desirable goal, it is far from worth the effort. The primary objective should be to have the data as accurate as needed in order that the product flow can be properly planned. This implies that management should focus its attention on improving the data that affect the synchronous operation of the system. For example, accurate bills of material, routings, and inventory files are critical to getting the product out the door on time. Time standards at CCRs must also be complete and accurate. But time standards at non-bottlenecks and non-CCRs are not as critical and therefore require less accuracy. The benefits of improving the data accuracy at non-CCRs will prove to be relatively minor.

A SYSTEMATIC APPROACH TO IMPROVING PERFORMANCE

A primary task of management is to continually improve the operation by reducing the restrictive effect that constraints have on the firm's ability to make money. The discussion to this point in the chapter has provided a sufficient understanding of constraints so that a logical approach to improving a firm's performance may now be presented. The recommended approach to improving any system includes the following steps: [34]

1. Identify the constraints in the system.
2. Determine how to exploit the constraints to improve the performance of the system.
3. Subordinate all parts of the manufacturing system to the support of step 2.
4. Carry out the steps necessary to improve the performance of the system.
5. If, in the previous step, a constraint has been broken or a new constraint develops, go back to step 1.

This process may be illustrated with a simple example. Consider a plant that contains a focused factory which produces only two products, identified as product C and product D. The selling price, material cost, labor requirements, potential market demand, and labor availability for each product are indicated in Table 4.2.

TABLE 4.2   WHAT PRODUCT MIX SHOULD YOU PRODUCE?

	PRODUCT C	PRODUCT D
Selling Price	$90	$100
Material Cost	$45	$ 40
Labor Required per Unit	55 Minutes	50 Minutes
Market Demand	Unlimited	Unlimited
Total Available Labor Hours for the Focused Factory	160 Hours per Week	

Many marketing or sales managers make decisions with exactly this type of information. Based on the information contained in Table 4.2, what should the product mix be? Market demand is such that the plant can sell all that they can currently produce. Therefore, since product D has a higher selling price, a lower material cost, and requires less labor than product C, product D appears to be the obvious choice.

Now let's take a closer look at how products C and D are produced and evaluate the decision to produce product D exclusively. Additional information about the process is presented in Figure 4.4. In the figure, the product routings and the processing time requirements at each of four resources (R1, R2, R3, and R4) are identified. Material is processed through resources R1 and R3 to form component X. Material is processed through resources R2 and R3 to form component Y. And material is processed through resources R1 and R2 to form component Z. Product C is produced by assembling components X and Y with an additional piece of material at resource R4, while product D is produced by assembling components Y and Z with an additional piece of material at resource R4. It is also determined that each of the four resources is available for a total of 40 working hours each week.

If the weekly operating expense for this process is $5,000 per week, what is the expected weekly profit of producing product D according to our previous decision? To help answer this question, consider Table 4.3, which summarizes the amount of processing time at each resource that is required to produce one unit of each product.

FIGURE 4.4

**FIGURE 4.4   WHAT SHOULD YOU PRODUCE?
USING ROUTINGS AND PROCESSING
INFORMATION TO HELP IDENTIFY
CONSTRAINTS AND MAKE DECISIONS**

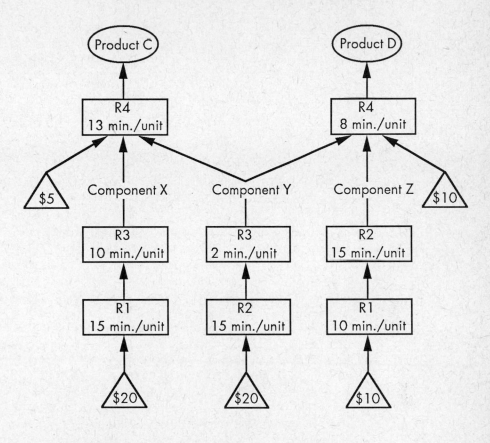

Key: ⟨$x⟩ = Material Costs

Since resource capacity appears to limit throughput in this situation, it is appropriate to identify the capacity constraint. An interesting point here is that if market demand is virtually unlimited, then by definition all four resources are bottlenecks. However, it is clear from the data in Table 4.3 that resource R2 is the only capacity constraint resource and is the only resource that must be considered in determining how much product can be produced.

TABLE 4.3 **WHAT SHOULD YOU PRODUCE?
ANALYZING PRODUCTION CAPABILITIES**

	PRODUCT C		**PRODUCT D**	
	(1)	(2)	(3)	(4)
	Processing	Maximum	Processing	Maximum
	Time per	Resource Output	Time per	Resource Output
RESOURCE	Unit (min.)	= 2400 min. / (1)	Unit (min.)	= 2400 min. / (3)
R1	15	160	10	240
R2	15	160	30	80
R3	12	200	2	1200
R4	13	184	8	300

Now consider the product mix in light of this new information. The throughput is limited to what can be processed by resource R2. Product D requires 30 minutes of processing time at R2 per unit. And since there are 2,400 minutes per week available at R2, only 80 units of product D can be produced each week. This level of production of product D yields weekly net product income (after material costs) for the firm equal to $4,800 (80 units per week × $60 contribution to profit per unit). With weekly operating expenses of $5,000, the decision to produce product D will result in a net loss of $200 per week.

Consider what happens if management only looks at the total amount of resource time available each week for all four resources (9,600 minutes) and divides that by the total amount of resource time required to produce a unit of product D (50 minutes). The likely, although incorrect, conclusion is that the process might be expected to produce as many as 192 units of product D each week. The actual production totals will be far below what management might expect. Moreover, the utilizations at resources R3 and R4 are extremely low because they can only work on material that has already gone through the bottleneck. In addition, the only way that R1 can have a high utilization is if it is overactivated and feeds excess material into the system.

What happens if product C is produced instead of product D? Since each unit of product C requires only 15 minutes of resource R2, it is possible to produce 160 units of product C each week. (If only product C is produced, then resource R1 also becomes a CCR, since it requires as much time to process a unit of C as resource R2.) The production of 160 units of product C results in weekly net product income (after material costs) of $7,200 (160 units per week × $45 contribution to profit per unit). After subtracting the

$5,000 weekly operating expenses, the firm stands to make a net profit of $2,200 per week from this operation if only product C is produced.

This illustration demonstrates the first two steps in the five-step approach. In order to consistently make the right choices, it is first necessary to identify the constraints in the system. In this case, the primary constraint is resource R2. To develop the best possible strategy, the constraints must be exploited. In the above example, exploiting the constraints means choosing the product mix that best utilizes the available time at the CCR. As we have seen, choosing product C over product D is the difference between making a healthy profit and losing money.

To execute step 3 in the five-step approach, the entire system must be subordinated to totally support the strategy identified in step 2. The strategy is to produce 160 units of product C each week. To support this, exactly 160 units of components X and Y must be produced every week. Furthermore, only materials needed to produce these required component parts should be released into the system. Anything less will artificially limit throughput. Anything more will create excess inventory in the system. An additional important point is that all of the resources should be evaluated based on how well their actions support the selected strategy and the desired product mix.

Step 4 requires that management take steps to improve the system. This can be accomplished in many different ways. Suppose in this example that management decides to purchase a new piece of equipment at resource R2 that triples the processing capacity at R2. This cuts processing time per unit at R2 to 5 minutes for C and 10 minutes for D. This breaks the bottleneck at R2 and allows for an increased level of throughput for the entire system. To determine whether the acquisition of the new machine actually improves the system, the purchase cost must be weighed against the increased revenues generated by the additional throughput. Determining the cost/benefit of the proposed change actually leads to the fifth and last step in the procedure.

Step 5 says that if a constraint is broken or a new constraint develops, then we should start over again back at step 1. That is, go back and identify the constraints in the system. This is a very critical point in the process, and one that is easy to overlook. The point here is that nothing should be assumed about the way the system will operate after a change in the system occurs. Every aspect of the system should be carefully reevaluated.

To illustrate step 5, since resource R2 is no longer the primary constraint in the system, something else must be the limiting constraint. In this case, resource R1 becomes the most constraining resource, essentially determining what the factory can produce. Since R1 requires 15 minutes to process a unit of C and 10 minutes to process a unit of D, now the plant can produce either 160 units of C or 240 units of D.

If management ignores step 5 and does not reevaluate the total system, a critical error may be made. For example, what if management assumes that product C is still the most profitable item to produce? According to the limitations of resource R1, only 160 units of product C can be produced, exactly the same quantity as before the equipment purchase for resource R2. Therefore, if the product mix (produce only C) is unchanged, then throughput remains unchanged, and the purchase of new equipment results only in a wasted expense.

Let's now consider that additional constraints may surface. For example, it was previously assumed that the plant could sell all of any item that could be produced. Now that the capacity to produce has been increased, is this still true? What if closer scrutiny of the market indicates that the plant can sell no more than 300 units of C and 200 units of D per week. This information was not previously significant because the plant could not produce those quantities of products. But now that the plant has increased capacity, market demand for product D becomes a constraint.

As previously mentioned, the plant now has the capacity to produce either 160 units of C, 240 units of D, or some combination of C and D. Furthermore, market demand now limits possible throughput to 300 units of C and 200 units of D per week. If only product C is produced, net profit remains at $2,200 per week as previously calculated. If only D is produced, 200 units can be sold, resulting in a net product income of $12,000 per week (200 units per week × $60 contribution to profit per unit). Net profit is $7,000 per week after subtracting the $5,000 weekly operating expenses. However, after 200 units of product D are produced, there are still 400 minutes of unutilized processing time at resource R1 each week. This is sufficient capacity to produce 26 units of C. Therefore, the best product mix would be to produce 200 units of D and 26 units of C each week. This yields a net profit of $8,070 per week. Clearly, if management ignores step 5 and does not consider that the product mix should change, then the plant will lose the difference between $8,070 and $2,200, or $5,870 per week.

This illustration demonstrates that the five-step approach to improving the performance of the firm is very powerful. Managers should thoroughly familiarize themselves with this procedure so that it becomes second nature. This approach is widely applicable to all parts of the firm and promotes the development of a synchronous manufacturing system.

SUMMARY

Constraints exist in every organization and effectively limit the ability of the firm to improve their productivity and make more money. Constraints may appear in many forms. The market may be a major limitation on what the firm can achieve. Material and capacity constraints place physical limitations on what the firm can produce. Logistical, managerial, and behavioral constraints introduce inefficiencies into the organization and may also aggravate the problems caused by the market and physical constraints.

Capacity constraint resources (CCRs) are those resources which, if not properly scheduled and managed, are likely to cause significant problems in the planned product flow through the plant. CCRs must be a focal point of managerial attention and may be either bottleneck or non-bottleneck resources. The identification and management of CCRs are a major consideration in the implementation of synchronous manufacturing in a plant.

The five-step approach first developed by Goldratt emphasizes the fact that there are only a few constraints that primarily limit the firm's performance at any given time. This approach also provides a basic structure that can be used to help implement synchronous manufacturing in a plant.

QUESTIONS

1. Why must every organization have at least one constraint?
2. Give an example of a logistical, managerial, or behavioral constraint not mentioned in the text. Explain how the identifed constraint restricts performance in the organization.
3. Consider a firm that produces only one product. Is it likely that this firm will simultaneously have market, capacity, and material constraints? Explain your answer.
4. Consider a firm that produces only one product. Is it possible that this firm will simultaneously have logistical, managerial, and behavioral constraints? Why or why not?
5. In a manufacturing plant, is it inevitable that the actual product flow will deviate from the planned flow? Explain your answer.
6. Explain why a CCR may not necessarily be a bottleneck.
7. Explain how bottlenecks and CCRs affect the product flow in a plant.
8. When management successfully breaks a constraint, what is likely to happen next, and what should management do about it?

PROBLEMS

1. Refer back to Table 4.1 in the text. Suppose the demand for product A increases to 3 units per day and the demand for product B decreases to 4 units per day. Indicate whether each of the four resources is a bottleneck and explain why. Also indicate whether each resource is a CCR and explain why or why not.
2. This table shows that two products, A and B, have the same routing. Demand for product A is 10 units per day and for product B is 20 units per day. Assuming that available capacity is 480 minutes for each resource, classify resources R1, R2, R3, and R4 as bottleneck, non-bottleneck, CCR, and non-CCR.

Operation	Resource	Product A Run Time per Unit	Product B Run Time per Unit
010	R1	15 minutes	15 minutes
020	R2	20 minutes	20 minutes
030	R3	30 minutes	5 minutes
040	R4	18 minutes	10 minutes

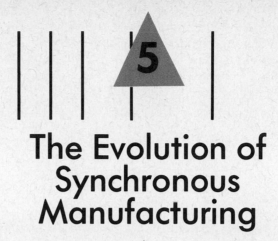

The Evolution of Synchronous Manufacturing

THE CONCEPT OF WASTE

In 1921, Henry Ford wrote in his book, *Today and Tomorrow*, "Having a stock of raw materials or finished goods in excess of requirements is waste." [10, p. 108] However, Ford's writings and actions indicate that his definition of what constituted material waste included all categories of inventory stocks, not just raw materials and finished goods. In the entire Ford industrial complex of Henry Ford's day, not a single warehouse was owned or used for the storage of materials, component parts, subassemblies, or finished goods. It is evident that Henry Ford's concept of waste reduction is totally consistent with the principle of synchronized material flows. Henry Ford was certainly a man of great insight and vision. And he was undoubtedly the first industrialist to truly understand and apply many of the basic principles inherent in the philosophy of synchronous manufacturing.

In most successful organizations, the elimination of waste is a prime objective. The problem has always been how best to achieve this objective. In order to logically approach this problem, the concept of waste must be clearly understood. The following is a definition of waste in the manufacturing environment:

Waste—Any activity in an organization that does not contribute to the common companywide goal of making money.

Waste within a manufacturing environment is anything in excess of the minimum amounts of materials, parts, equipment, facilities, labor, and time that are essential to produce and deliver the desired product to the customer. The key to the elimination of waste in a manufacturing firm lies in management's ability to develop and maintain a synchronized flow of materials and products into, through, and out of the plant in concert with market demand. The degree to which the material and product flow of a plant can be synchronized determines the degree to which waste can be eliminated and, therefore, the degree to which the plant can be competitive in the marketplace.

LESSONS FROM HISTORY

History provides many great lessons if we will only take the time to reflect. An analysis of the strengths and weaknesses of the assembly line and the just-in-time logistical processes is conducted later in this chapter. But first, a brief overview and comparison of the management philosophies of Henry Ford and progressive Japanese managers such as those at Toyota Motor Company is presented.

The Ford System

Henry Ford recognized the importance of eliminating waste within the manufacturing environment and designed the Ford industrial structure accordingly. Each of the following statements is taken directly from Ford's book, *Today and Tomorrow*. [10] These statements provide an interesting glimpse of the Ford industrial organization and Henry Ford's philosophy of management and productivity.

Henry Ford on the advantage of a fully integrated manufacturing system:

> It was in order to eliminate lost motion—which is just as fatal in a factory as in a bearing—that we began, some years ago, the plant which we call Fordson and which has become the heart of our industries. . . . The plant covers more than a thousand acres, has a mile of river frontage, and employs upward of 70,000 men. (p. 101)

> It is inevitable that the business of the country shall be done by very large companies which reach back to the source, and, taking the raw material, carry it through the necessary processes to the finished state. Just as soon as a business gets beyond a certain size the control of materials has to be absolute, for, regardless of costs, it is not otherwise possible to avoid the stoppages of strikes or the advent of unskilled management. (p. 244)

> If Fordson did not deal in heavy, bulky raw materials it would not pay. It pays because it combines quick transportation both inward and outward. As a general rule, a large plant is not economical. (p. 109)

Henry Ford on the importance of maintaining the product flow:

> The thing is to keep everything in motion and take the work to the man and not the man to the work. That is the real principle of our production, and conveyors are only one of many means to an end. (p. 100)

> The men do not leave their work to get tools—new tools are brought to them . . . machines do not often break down, for there is continuous cleaning and repair work on every bit of machinery in the place. (p. 100)

Henry Ford on the necessity of good material control:

> Last year we made certain changes to the end of turning out somewhat better cars. . . . We set a date to begin changing over. The planning department had to calculate on just the amount of material which would keep production going at full speed until that date and then permit production to stop without having any material over. (p. 87)

> The problem of coordination is simplified by standard carloads. (p. 114)

> We want to use material to its utmost in order that the time of men may not be lost. Material costs nothing. It is of no account until it comes into the hands of management. We will use material more carefully if we think of it as labor. (p. 91)

> Time waste differs from material waste in that there can be no salvage. On the other hand, it is a waste to carry so small a stock of materials that an accident will tie up production. The balance has to be found. . . . (p. 110)

Henry Ford on the role of quality and inspection:

> The key to our production is inspection. More than three percent of our forces are inspectors. This simplifies management. Every part in every stage of its production is inspected. (p. 100)

> We must have inspectors at every stage of the work; otherwise, faulty parts might get into the assembly. Our inspectors in only a few cases are required to use judgement. . . . (p. 77)

Henry Ford on the role of labor:

> Our system of management is not a system at all; it consists of planning the methods of doing the work as well as the work. All that we ask of the men is that they do the work which is set before them. (p. 100)

In summary, Ford concentrated his efforts on eliminating waste in all aspects of the operation. The flow from raw materials to finished product was streamlined. Strict material planning and control procedures were instituted. Procedures were developed to try to minimize lost motion and lost production time. Continuous preventive maintenance was performed.

Quality inspections were done continuously at every step of the operation in an effort to produce a high-quality product. Workers were required to perform only the work that was placed in front of them. Significantly, local efficiency measures were not used to evaluate worker performance. The ultimate measure was the overall performance of the system.

One could argue that, in terms of what was actually accomplished, Ford was the world's first large just-in-time manufacturer of discrete goods. In all parts of the Ford industrial complex, materials arrived just in time for processing; semifinished parts arrived just in time for machining and fabrication; and finished parts and subassemblies arrived just in time for final assembly.

The Japanese Just-in-Time System

The highly successful Japanese just-in-time (JIT) manufacturers, led by Toyota, learned many valuable lessons from Henry Ford. Many of the cornerstone principles of the Japanese JIT system were first used by Ford 70 years ago. The reader who is familiar with the JIT philosophy will recognize that many of the Ford management principles stated above are indeed key concepts in JIT systems. The Japanese have also modified and improved many of the Ford principles as they developed their JIT system. [11, 13, 14, 15]

As with Ford, the elimination of waste is a fundamental driving force of JIT philosophy. This elimination of waste philosophy fits well with Japanese society. The population of Japan, approximately 130 million, is about half that of the United States. Yet the islands of Japan constitute a total land mass about the size of the state of Montana. Moreover, much of the terrain is rugged, mountainous, not very habitable, and significantly lacking in valuable natural resources. As a matter of necessity, the Japanese have learned how to avoid wasteful practices and how to compete successfully with limited resources.

The Japanese have developed many techniques to expose inefficiencies in all phases of their operation in order to gain opportunities to eliminate waste. The JIT philosophy has guided the development of systems designed to reduce inventory, scrap, rework, and the space and equipment required to handle inventory, while improving product quality and market responsiveness.

One major development in the Japanese JIT system is the expanded role and responsibility given to the worker. The workers are not treated as extensions of a machine. The Japanese workers are well educated and well trained. They are given a tremendous amount of responsibility to maintain their equipment, check for quality problems, develop new ways to improve the process, and act as problem solvers.

Later in this chapter, a more in-depth discussion of the assembly line and JIT philosophies is conducted. At that point, the degree to which these two philosophies adhere to and violate the principles of synchronous manufacturing are examined.

What Went Wrong?

Seventy years ago, Henry Ford demonstrated how to set up and run a well-synchronized production facility. The Japanese learned well from Henry Ford and applied many of his management principles to their own organizations. The burning question is, "What happened in this country?" How have domestic firms managed to become so uncompetitive in the international marketplace? No single answer can completely explain this phenomenon. However, the fundamental explanation is clear.

After World War II, the manufacturing capacity of U.S. firms dominated the world marketplace. The demand for products was great and domestic manufacturers had little competition. Such conditions naturally encouraged wasteful practices and allowed unsound managerial policies to evolve slowly. The focus of many organizations shifted to meeting market demand at any cost. Production costs were of little consequence since there was not much competitive pressure and prices could usually be raised to cover increasing costs.

Eventually, worldwide manufacturing capacity grew to the point where maintaining a strong competitive position became a critical issue. In response, management became increasingly reliant upon cost control systems in an effort to remain competitive. Ultimately, the attainment of production efficiencies and local optima were given top priority in the race to remain competitive. Managers of large, complex organizations relied upon their cost-based systems to provide the correct answers to critical operational decisions. In the final analysis, what happened is that the managers of U.S. firms forgot the lessons of Henry Ford and unwittingly shifted their focus away from synchronized production flows.

THE IMPORTANCE OF PRODUCT FLOW

In Chapter 2, the concept of a synchronized product flow was discussed relative to its effect on the competitive edge factors. The significance of product flows in attaining a synchronous manufacturing environment will now be further developed.

Time and Product Value

According to Henry Ford, "The time element in manufacturing stretches from the moment the raw material is separated from the earth to the moment when the finished product is delivered to the ultimate customer." [10, p. 108] Henry Ford seemed quite proud of the fact that, in 1921, the company's production cycle from the mine to the finished automobile on the freight car was approximately 81 hours.

How many manufacturing firms today, including Ford Motor Company, would not be thrilled with a total manufacturing cycle of 81 days, much less 81 hours? At the same time, Henry Ford proclaimed ". . . we shall never reach the point where whatever we happen to be doing cannot be improved upon." [10, p. 242] Our managers must realize that time is something that erodes the product value. That is, the greater the amount of time that passes in the production and sale of a product, the less is the value of that which has been produced.

The Role of Product Mix

It is interesting to note that the 81 hour manufacturing cycle time equals the best performances of Japanese JIT auto manufacturing plants today. [15, p. 7] How was it possible for Ford to achieve such a remarkable feat in 1921? Part of the answer lies in his famous statement, "They can have it in any color they want, so long as it's black."

The most efficient form of production is a continuous flow process like those found in refining, bottling, or extrusion processes. The products produced within each batch or run in these types of processes are essentially identical in nature. Ford clearly recognized the manufacturing economies and waste-reduction possibilities in these types of processes and designed his products and plants accordingly. Ford's facilities were essentially dedicated plants and production lines. These plants mass produced all of the interchangeable components and subassemblies necessary to build the final product. These facilities were very economical since all units produced were identical. It isn't easy to look like a continuous flow producer when manufacturing discrete products, but Ford came close. One can almost imagine a complex "pipeline" network at Ford, through which the materials and components "flowed" to the right place at the right time in just the right quantity.

Compared to the Ford Motor Company of 1921, the Japanese today offer a universe of product models and options. However, Japanese plants are typically more focused or dedicated in their product lines than similar manufacturing firms in the United States. They simply standardize and limit the number of variations on the basic product line that they produce. But this focused factory concept is only partially responsible for the manufacturing advantage enjoyed by the Japanese in today's market. Their success is primarily due to the way in which they view the overall purchasing/manufacturing/ marketing process.

A River Analogy

One of the key concepts in synchronous manufacturing, which the Japanese understand very well, is that of smooth and timely material flows into, through, and out of the plant. The objective is that raw materials, components, work in process, or finished goods will not be stagnant at any point in the process from receiving to shipping.

It is enlightening to compare the concept of material flow in a manufacturing environment to a river. Envision a river flowing through the countryside. The river, as shown in Figure 5.1, is of varying depths with occasional deep pools. The speed of the flow in the river also changes, depending on the width and depth of the river and the change in elevation of the surrounding landscape. Within the river, there are large boulders, submerged trees, and other obstacles that impede and divert the flow. Now relate the river environment to that of a typical manufacturing environment. The flow of water in the river represents the flow of materials through the plant. The depth of the water is symbolic of the amount of inventory in the system. The deep pools represent piles of stagnant inventory. The boulders, trees, and other obstacles in the river symbolize the many problems in a manufacturing environment that disrupt the production process and impede the smooth flow of materials through the system. [16, p. 17]

The ideal production process has several characteristics that can be symbolized within the river analogy. First, the flow of materials through the manufacturing system should be constant and in concert with market demand. Alternating periods of floods and droughts are as devastating in a manufacturing plant as they are in a river basin. Second, the level of inventory

FIGURE 5.1   THE FLOW OF A RIVER

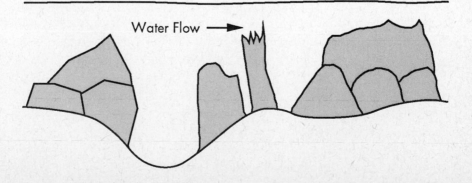

Water Flow →

in the plant should be reduced, and the piles of stagnant inventory should be eliminated. This is equivalent to lowering the water level of the river (while keeping the total flow constant) and eliminating the deep pools. Third, the disruptions within the manufacturing environment must be eliminated in order to achieve a smooth production flow. In the river, the boulders, submerged trees, and other obstacles must be removed.

To continue the analogy, consider a boat traveling on the river. Trying to steer a boat successfully down a river filled with obstructions is difficult. Trying to run efficiently a manufacturing plant that is characterized by significant disruptions is equally difficult. If the river is to be successfully and safely navigated, then all obstacles and hazards should be removed. Of course, the obstacles must be identified before they can be removed. In the manufacturing plant, this is equivalent to correcting the disruptions that prevent an efficient production process. The obvious problems are the easiest to solve and in many plants have already been eliminated. In the river, even if a few visible obstacles still exist, a course can be set so that the boat may be steered around them. The real dangers to safe navigation are the submerged hazards that cannot be seen. Similarly, in a manufacturing plant, the real threats to a smooth, successful production process are the unseen or unrecognized disruptions and problems that continually plague the manufacturing environment. Figure 5.2 illustrates some of the problems that abound in most plants and drive many managers to early retirement.

The traditional approach to alleviating the effects of disruptions in manufacturing facilities has been to keep extra stocks of inventory, "just in

FIGURE 5.2   OBSTACLES TO THE SMOOTH FLOW OF MATERIALS THROUGH THE PROCESS

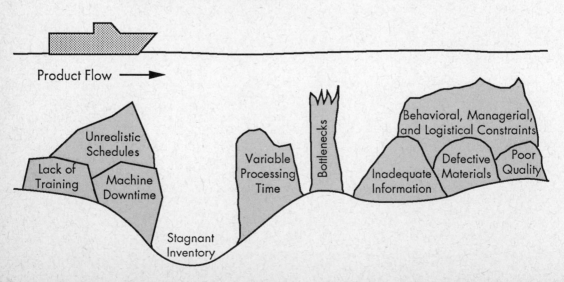

case" there is a big problem. As demonstrated in Chapter 2, large stocks of inventory, particularly work-in-process inventory, are not consistent with a synchronized flow philosophy and destroy a firm's competitive edge. Figure 5.3 illustrates that "just-in-case" inventory is used not to solve the problems and make the process more productive, but to cover up the problems so their effects are not directly felt. Figure 5.3 shows a raised water level (representing the "just-in-case" inventory), which prevents the boat from crashing into submerged boulders. This approach may give managers a feeling of great security. But meanwhile, the nonsynchronized material flows along with the hidden problems, and disruptions are killing the plant's productivity and ability to compete.

The just-in-time approach is quite different. Instead of increasing the inventory level (raising the water level) to cover up the problems, the inventory is reduced (the water level is lowered) so that problems can be more easily identified. Once the problems are uncovered, then the solutions are actively pursued. Figure 5.4 shows the disadvantage of this approach. In the figure, the boat has foundered upon a huge submerged boulder. Reducing the inventory has succeeded in identifying a problem; but that problem has temporarily scuttled the production process.

A primary objective is to improve the manufacturing environment so that a rapid, constant, and smooth flow of materials through the process is realized. This is certainly not accomplished by increasing inventory levels in an effort to cover up the problems. Nor is it necessary to lower the inventory levels

FIGURE 5.3   "JUST-IN-CASE" INVENTORY COVERS UP PROBLEMS

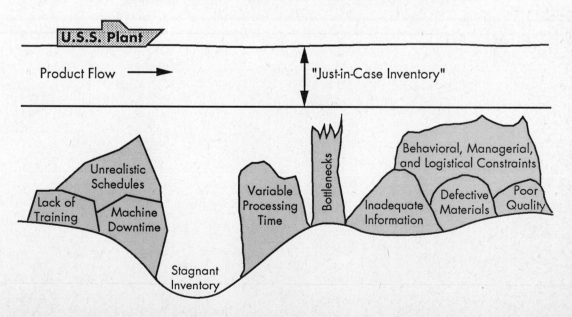

FIGURE 5.4   JUST-IN-TIME APPROACH UNCOVERS PROBLEMS

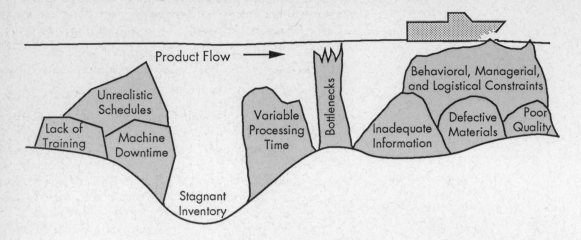

recklessly in order to locate the problems more easily. Instead, there is another alternative. Why not develop a system that, like underwater sonar, will clearly identify the problems that disrupt the flow? There is no need to dash headfirst into the rocks in the effort to find them. The objective should be to locate, identify, and remove the problems without causing serious harm to the production process. Chapter 6 presents a logistical system that allows us to use a sonar-like procedure to pinpoint the most significant disruptions to the material flow through the plant.

The discussion of constraints in the previous chapter should serve to remind us that the boulders that disrupt our plants are not only physical boulders. They may be managerial, behavioral, and logistical boulders as well. These are counterproductive policies and practices that cause or magnify inefficiencies in the production process and usually result in excess inventory. These nonphysical boulders can be just as harmful as physical boulders. A classic example of a managerial boulder that has shipwrecked many companies is the use of performance evaluation systems that encourage dysfunctional behaviors. A more recognizable managerial boulder is the use of batch sizes that are too large. Large batch sizes prevent the development of a smooth, continuous flow of materials through the system, making it impossible to establish a synchronized process.

THE CONCEPT OF BATCH SIZE

It is difficult to overestimate the importance of the batch-sizing decision in a manufacturing environment. Improperly chosen batch sizes contribute to nonsynchronized material flows. The direct result is increased levels of inventory and operating expense. The timely production of throughput may also be negatively affected. But the ultimate result is likely to be loss of market share, late shipments, inferior quality, excess investment in plant and equipment, decreased profit margins, and overall poor plant performance.

It is critical that batch-size decisions be made in the context of what is best from the perspective of total firm performance. Unfortunately, the traditional cost-based batch-sizing rules such as economic order quantity only serve to divert our attention and lead us astray. All cost-based batch-size approaches attempt to attain local optima and, accordingly, are typically suboptimum for the global (companywide) system.

Batch-size decisions must be viewed from the perspective of what effect they have on the synchronous flow of materials and products through the plant. Batch-size decisions that are not consistent with the synchronous manufacturing philosophy are disruptive to a synchronous flow and are not optimum. In the following sections, it is clearly demonstrated that the traditional cost-based approaches to batch-size determination are neither valid nor consistent with the principles of synchronous manufacturing. [32]

Traditional Approach to Batch-Size Decisions

In the traditional approach to batch-size decisions, two factors weigh heavily in the analysis. These factors are the total annual setup cost and the total annual carrying cost. There is an inverse relationship between batch size and setup cost. As the batch size increases, the number of batches that must be processed over the period of a year decreases, resulting in fewer setups and a lower total annual cost for setups. The relationship between batch size and carrying cost is a direct one. The larger the batch size, the higher the total annual carrying cost. This relationship is based on the approximately correct premise that as batch size doubles, work-in-process inventory doubles. And traditional cost concepts tell us that if the average level of inventory doubles, then total annual cost to carry that inventory doubles.

Since total annual setup cost and total annual carrying cost act in opposite directions, the usual approach is to establish the optimum batch size as that quantity for which the sum of these two costs are minimized. This optimum batch-size quantity is often referred to as the economic order quantity (EOQ) or economic batch size. Figure 5.5 illustrates the total cost curve as a function of the batch size and identifies the EOQ point. It is the EOQ and its derivatives that have guided our decision-making processes for years. It is now time to understand why these optimum batch sizes are totally invalid.

Invalid Assumptions of the Traditional Approach

The whole class of EOQ batch-sizing rules are based on three very erroneous assumptions. First, the analysis assumes that the carrying cost

FIGURE 5.5 TOTAL COST CURVE FOR EOQ BATCH-SIZING APPROACH

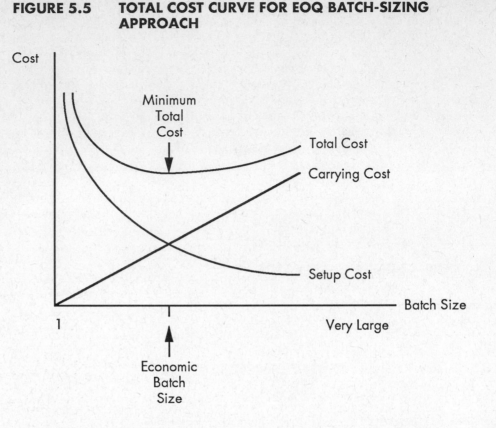

comprises only the dollar cost to actually carry the inventory. Second, the analysis does not make a distinction between the value of setup time at a bottleneck versus a non-bottleneck resource. Third, the entire traditional approach to batch-sizing decisions assumes that there is only one aspect of batch size to consider when, in fact, there are two distinct and very different types of batches.

The Carrying Cost Calculation The traditional approach to the calculation of carrying cost includes only the typical cost items such as the cost of capital, storage and handling cost, insurance, taxes, obsolescence, and deterioration. However, this approach to calculating the cost of carrying work-in-process inventory grossly understates the true cost. Think back to the discussion of how excess inventory adversely affects a firm's lead time, product quality, delivery performance, and product cost. It can be argued that the extra carrying cost incurred for excess work-in-process inventories is insignificant compared to their catastrophic effect on the firm's competitive edge.

No Distinction between Resource Types   In all of the traditional approaches to batch sizing, no reference is ever made to whether or not the resource is a bottleneck. The implicit assumption is that it makes no difference whether a resource is a bottleneck or not. In reality, it makes all the difference in the world. The value of an hour of processing time at a bottleneck is equal to the net value of the throughput for the entire plant. But the marginal value of an hour at a non-bottleneck is zero since the resource has excess capacity. Thus, the cost of setup time depends largely on whether or not the resource is a bottleneck. Since the cost approach inherent within the EOQ analysis does not make this distinction, it ignores the realities of the manufacturing environment, and the conclusions must be considered invalid.

The Single Batch Concept   The traditional approach assumes that batch size is a single entity. But there are actually two aspects of batch size that must be considered. The assembly line environment provides an excellent framework within which to examine these two aspects of batch size.

In a dedicated assembly line process that produces only one standardized product, what is the batch size? Some would answer that it is one unit, while others would say that it is the size of the production run—a very large number. Actually, both answers are correct!

In the dedicated assembly line, there are no setups. There is just one continuous batch. As a result, from the *resource* perspective, the batch size is the total number of units produced during the production run. But to keep the product flow constantly moving, as each work station finishes its work on a unit, that unit is quickly moved to the next station. From the *product* perspective, each unit is processed and moved individually, and so the batch size appears to be one unit. These two observations about the batch size are contradictory, yet both are valid. The conclusion must be that there is not just one but actually two distinct types of batches.

Process and Transfer Batches

The two different types of batches are referred to as the process batch and the transfer batch. The following are definitions of these two terms:

Process Batch—The quantity of a product processed at a resource before that resource changes over to produce a different product.

Transfer Batch—The quantity of units that are moved at the same time from one resource to the next.

It is clear that the setup cost at any resource should be a primary determinant of the process batch quantity at that resource. In the dedicated assembly line, setup costs are very high, causing the resulting process batch to be very large. Conversely, the cost of carrying inventory (including consideration of the effect on the competitive edge factors) affects the transfer batch quantity.

In the dedicated assembly line, the carrying cost is high and carrying excess inventory is extremely inefficient. Thus, the transfer batch is one unit.

The dedicated assembly line is a most efficient processing environment for discrete products. These observations lead to the sixth principle of synchronous manufacturing: [32]

Synchronous Manufacturing Principle 6: The transfer batch need not, and many times should not, equal the process batch.

Ideally, to help maintain a synchronous flow in a manufacturing process, the transfer batches should be small. At non-bottleneck resources, the process batches should generally be kept as small as possible in order to keep the flow of materials smooth and balanced. But at bottleneck resources, the small transfer batches must be combined into relatively large, economical process batches.

Keep in mind that manufacturing environments are dynamic and full of complex interactions and statistical fluctuations. Furthermore, the effects of the schedule, routings, priorities, and other constraints and considerations within the system will require that the size of the process batch change over time. Even for a given product at a given resource, the optimum process batch will almost certainly be variable over time. These considerations lead to the seventh principle of synchronous manufacturing: [32]

Synchronous Manufacturing Principle 7: A process batch may be variable both along its route and over time.

EVALUATING ASSEMBLY LINES AND JUST-IN-TIME SYSTEMS

In Chapter 2, the adverse effects of an unsynchronized flow on the firm's competitive edge was clearly demonstrated. In the troop analogy of Chapter 3, the spreading-of-the-troops phenomenon was used to symbolize the natural propensity of the material flow to become nonsynchronized in a manufacturing environment.

Our attention now turns to an evaluation of assembly lines and just-in-time systems. The objectives of this section are threefold. First, to develop an understanding of how assembly lines and JIT logistical systems work. Second, to determine the degree to which these two systems apply the principles of synchronous manufacturing. Third, to analyze the limitations of these two systems, especially in view of the principles of synchronous manufacturing.

Assembly Lines

To further develop our understanding of assembly lines, we will discuss how and why well-designed assembly lines work. We will also identify and explain their limitations. A good way to begin our evaluation of assembly lines is from the perspective of the troop analogy presented in Chapter 3.

The Troop Analogy Revisited   One way to prevent the spreading of the troops is to tie the rows of soldiers together physically with ropes. [5, pp. 88–89] Mountain climbers often use this technique. (A sudden large spreading between mountain climbers can have quite serious consequences and is a phenomenon to be avoided.) Figure 5.6 illustrates this approach to the problem. The ropes guarantee that the movement of the troop is relatively synchronized. The speed of the troop is limited to the speed of the slowest soldier. The ropes also limit the amount of spreading that could occur within the troop. The shorter the ropes, the less the amount of spreading.

Physical Ropes for Assembly Lines   In order to apply the concept of physical ropes to a manufacturing environment where the production resources are physically connected, three requirements must be met. First, the finished goods being produced must be fairly standardized. Second, the processes required to produce the finished goods must be such that there is a single-direction, in-line flow of material from one work station to the next. Third, there must be a high level of demand for the product in order to warrant special manufacturing lines dedicated to the high-volume production of a standardized product.

Ford's manufacturing process clearly met all three of these requirements. Thus, Henry Ford decided to apply the basic concept of physical ropes and developed his assembly line-based production facilities. He connected his

FIGURE 5.6   THE USE OF PHYSICAL ROPES TO PREVENT SPREADING OF THE TROOPS

assembly line production resources with a system of conveyor belts—a reasonably close logistical facsimile of physical ropes.

Whether or not Henry Ford initially realized all the ramifications of this logistical approach to manufacturing is not clear. But the end result was that the workers at Ford were forced to work as a synchronized unit. The pace of the assembly line was determined by the speed of the conveyor belt and the spacing between units on the belt. This pace, which could be adjusted to the speed of the slowest worker or work station, determined the overall production rate for the entire factory. Through this approach, Ford achieved previously unimagined success in mass production.

Ropes Limit WIP Inventory   The overall length of the conveyor belt in an assembly line process effectively limits the total amount of spreading, or work-in-process inventory, in the system. Likewise, the spacing between individual work stations along the line controls the amount of work-in-process inventory between stations. In addition, the size of an individual work station determines the total amount of time that a specific unit is available to be worked on at the station. By varying the size of the work stations and the spacing between stations, the work-in-process inventory and the material flow can be effectively controlled.

It is the tightly controlled availability of work in process at each station that makes the assembly line effective. The conveyor system dictates to the workers when to work and, just as important, when not to work. Thus, neither managers, supervisors, nor workers are able to introduce excess materials or work-in-process inventory into the system. As a result, the work along the line becomes synchronized, with every work station processing materials at the same rate.

Why Assembly Lines Work—Incorporating the Principles   Assembly lines have been successfully implemented in a limited number of manufacturing environments. Assembly line logistical systems are usually quite successful because such systems allow the development of highly synchronized material flows. We now examine the extent to which assembly lines adhere to the already established principles of synchronous manufacturing:

> *Principle 4:* The level of utilization of a non-bottleneck resource is controlled by other constraints within the system.
>
> *Principle 5:* Resources must be utilized, not simply activated.

Synchronous manufacturing principles 4 and 5 are strictly adhered to in most assembly line environments. Where the material flow is automatically paced and beyond the control of individual workers, the flow has an excellent chance to be synchronized. In this type of environment, individual workers are asked only to perform the work that is placed before them. Thus, it is impossible for individual workers to be activated without being utilized.

Furthermore, the level of utilization of each resource is determined by the speed of the line.

> *Principle 6:* The transfer batch may not, and many times should not, be equal to the process batch.

Synchronous manufacturing principle 6 was first developed and explained in terms of a dedicated assembly line environment. Obviously, since one characteristic of assembly lines is relatively small transfer batches coupled with relatively large process batches, this principle applies to all assembly lines.

Most assembly line environments fully incorporate synchronous manufacturing principles 4, 5, and 6. Thus, it is easy to understand the success of these systems in developing synchronized flow processes.

Limitations of Assembly Lines   Unfortunately, it is not feasible to rely upon assembly line logistical systems to help us attain the competitive edge. There are three major drawbacks to the use of assembly lines in our manufacturing firms. First, assembly lines are of limited applicability. Second, they require huge investments of both capital and labor to design, develop, set up, and balance. Third, the throughput of assembly line processes is highly vulnerable to the disruptions that often occur within the typical manufacturing environment. This vulnerability increases as the capacities of the various work stations on the line are balanced.

Limited Applicability   An assembly line is an extremely efficient manufacturing process that can be utilized only for fairly standardized products for which there is high-volume demand. Furthermore, since conveyor belts physically connect all the production resources, this logistical system can only be applied to those processes where each work station feeds work-in-process material to one, and only one, work station. This type of system clearly is not applicable to plants that process products characterized by dissimilar routings. Thus, the assembly line approach is automatically limited to a relatively small percentage of manufacturing industries since most plants' product lines are missing one or more of these necessary ingredients.

Required Level of Investment   The assembly line approach to manufacturing requires a dedicated processing line, usually consisting of a series of relatively expensive machines and material handling systems. This type of system typically requires a major effort to debug the line in order to achieve the necessary synchronized material flow. Closely associated with this long and arduous debugging effort is the obligatory attempt at line balancing. Line balancing is management's endeavor to develop approximately equal processing capacities at each work station along the line. In addition, change-overs from one product model to another can be extremely time-consuming and very expensive, and the debugging and line balancing procedures must start again from the beginning.

Effect of Disruptions to the Flow   In the column of marching soldiers—now physically connected by ropes—suppose one of the soldiers stumbles, falls to the ground, and spills his gear. After uttering a few well-chosen words, he eventually collects himself and rises to his feet to resume the march. But since the fallen soldier is physically tied to the rest of the troop, the column will be abruptly halted by this event. In a similar manner, the progress of the entire troop is impeded by every significant disruption that occurs during the duration of the march.

This situation can be applied to the manufacturing environment to illustrate vividly the major disadvantage of the assembly line logistical system. The primary drawback is the disruption to the flow that occurs whenever there is a significant problem at any single work station anywhere along the entire line.

In the traditional assembly line environment, a primary objective is to balance the line. In fact, the line balancing principles that are an integral part of assembly line activities violate synchronous manufacturing principle 1, restated here:

> ***Principle 1:*** Do not focus on balancing capacities, focus on synchronizing the flow.

When too much emphasis is placed on balancing the actual capacities of the resources with the production time requirements of the process, even small disruptions can create major problems for the entire process. Consider the extreme case where every work station on the line is perfectly balanced and equal to the work requirements. There is no excess capacity at any of the resources. Thus, even the smallest disruption anywhere along the line will quickly be felt throughout the entire process.

When a disruption occurs in a perfectly balanced line, there either will be a quality problem with the product, or the line will shut down. If the line does not shut down, product quality will suffer. If the line does shut down until the problem is fixed, then there is a loss of throughput. Neither of these alternatives is desirable. We are all aware of the negative effects of poor-quality products. Yet a shutdown of the line means lost throughput. A disruption that results in a shutdown is very expensive, since additional products could have been produced for only the marginal cost of the material.

This discussion illustrates why Henry Ford insisted on numerous procedures designed to reduce the disruptions to the assembly line product flow. His strict and continuous preventive maintenance procedures were designed to keep the machines from breaking down. His insistence upon inspection of every part at every stage of the operation was intended to identify and correct immediately problems that could disrupt the operation. In fact,

Ford was so concerned with the smooth flow of the operation that even paydays were designed to minimize disruptions. According to Ford, "Almost every hour of the day is payday somewhere in the plants." [10, p. 137] This saved workers lost hours waiting for their pay and also reduced the need for a large payroll staff.

Just-in-Time Systems

It should be recognized that JIT systems, especially those characterized by a logistical control system such as the Toyota kanban system, are not found in all types of manufacturing firms. Such JIT systems are predominantly found in firms involved in repetitive manufacturing.

Japanese Successes in Repetitive Manufacturing   Inventory turnover is a good measure of the degree to which the material flow in a manufacturing firm is synchronized. According to a study conducted by the consulting firm of Booz, Allen, and Hamilton, the inventory turnover rate for manufacturing firms during the 1970s averaged 5.7 in Japan versus 3.7 for firms in the United States. [35] Table 5.1 presents additional information from the study for job shop, repetitive, and process industries. According to the data in Table 5.1, there is little difference in the inventory turnover rates between Japanese and U.S. firms in the job shop and process industries. But Japanese firms involved in repetitive manufacturing averaged 7.6 turns, while repetitive manufacturers in the United States averaged only 3.8 turns.

The data shown in Table 5.1 contain another surprising result. The inventory turns for the repetitive manufacturing industries in Japan even exceeded the turnover rates for the Japanese process industries. Process industries, with their highly capital-intensive structure and continuous product

TABLE 5.1   A COMPARISON OF INVENTORY TURNOVER RATES FOR JAPANESE AND U.S. INDUSTRIES DURING THE 1970s

TYPE OF INDUSTRY	INVENTORY TURNOVER RATE	
	United States	Japan
Job Shop	3.2	3.1
Repetitive	3.8	7.6
Process	5.1	5.5
All	3.7	5.7

Source: Presentation by Bob Fox, 1981 APICS Conference

flows, have traditionally been considered the most efficient form of manufacturing. The evidence seems puzzling. But the conclusion must be that Japanese society and culture is not the primary reason for the Japanese success story, unless there is a different type of person working in the repetitive manufacturing industries.

The phenomenal success of the Japanese repetitive manufacturers is primarily due to the extensive application of just-in-time systems in these industries. The basic JIT concept is quite simple. Finished goods should be produced just in time to be sold, subassemblies should be completed just in time to be assembled into finished goods, fabricated parts should be processed just in time to go into subassemblies, and purchased components and materials should arrive just in time to be transformed into fabricated parts. The JIT ideal is that all material in the system be constantly flowing through the various transformation processes with batch sizes approaching one unit.

Logistical Ropes—How the JIT Kanban System Works   The troop analogy is again useful in illustrating how the kanban logistical system works. A modified approach to the rope system used by assembly lines can be applied to the repetitive manufacturers. For every different product produced by a manufacturing firm, a routing from the gateway operation to the finished product can be identified. Imagine a rope that runs from final assembly back through the process to the gateway work station for a single product. Now extend this concept so that every identifiable product has a rope connecting every work station contained in its routing.

In an assembly line environment, the ropes are conveyors—real physical entities. But how can the concept of physical ropes be extended to a manufacturing process that uses a variety of product routings to produce a number of different end items? The use of physical ropes, such as conveyors, cannot be used to connect the various resources in this type of environment. The resulting maze of material handling devices would be a true nightmare. Thus, another approach is necessary. The Japanese have developed what we refer to as "logistical ropes," which connect their various work stations and drive their JIT system.

The kanban system, first developed by Toyota, is the best known of these "logistical rope" systems used in JIT environments. In the typical kanban system, the master production schedule for the plant is determined according to market demand. The production schedule is set up in order to produce and ship some quantity of each item every day in concert with market demand. The projected average daily demand for each product is the amount that is scheduled to be produced each day. This smoothed production schedule is then set for a fixed period of time—normally 1 month. Then, final assembly (or some other final stage in the process) is scheduled in order to meet the daily production requirements.

In the Toyota system, which uses a dual-card kanban system, each work station has its own inbound and outbound material storage areas. The inbound stock area holds material that is ready for processing at the work station. In the outbound stock area, material that has already been processed by the work station is stored until it is needed at the next downstream station.

As production occurs according to the predetermined schedule, final assembly consumes standard containers of materials from its inbound stock areas. As these containers are taken from inbound stock, a signal is sent to the feeding work stations to replenish these materials from their outbound stock areas. In the Toyota system, the signaling is accomplished through the use of a conveyance kanban. This conveyance kanban is a card that identifies the needed material, the feeding work station, and the receiving work station. As the material is forwarded from the outbound stock areas, a card referred to as a production kanban is delivered to the feeding work station personnel. This production kanban is the signal and authority for the work station operators to replenish the material just forwarded from the outbound stock area. The work station operators then take material from its inbound stock areas to process and replenish its outbound stock as dictated by the production kanbans on a first-in, first-out basis. As the work station operators draw material from its inbound stock area, one cycle is complete. The signal to move and process work-in-process materials at other feeding work stations is passed down the line. Figure 5.7 illustrates this process for two component parts (A and B) as they move through the final stages of the process to final assembly.

This type of system is called a pull system. Only final assembly is scheduled, and the materials are pulled through the system only as needed by the downstream work stations. If a work station does not receive a signal (production kanban) to produce, no work is performed, even if plenty of materials are available and ready to be processed. It should be evident that the kanban system just described is really nothing more than a material replenishment system. The "logistical ropes" are used to notify the workers when materials need to be moved to replenish inbound stock and when materials need to be processed to replenish outbound stock.

Why Kanban Systems Work—Executing the Principles   The kanban concept works because the logistical system inherently recognizes, and at least partially applies, the basic principles of synchronous manufacturing that have been discussed up to this point. Exactly how these principles are executed is now examined more closely.

> *Principle 1:* Do not focus on balancing capacities, focus on synchronizing the flow.

Synchronous manufacturing principle 1 is an important aspect of kanban systems. The only formal predetermined schedule in a kanban system is for

FIGURE 5.7 **THE MOVEMENT OF PRODUCTS A AND B THROUGH FINISHING WORK STATIONS TO FINAL ASSEMBLY IN A DUAL-CARD KANBAN SYSTEM**

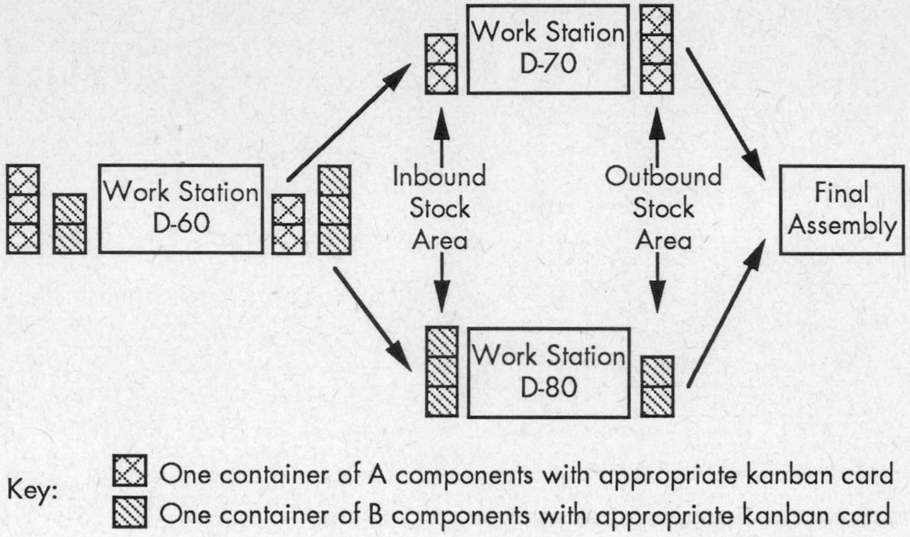

Key: ⊠ One container of A components with appropriate kanban card
      ⊡ One container of B components with appropriate kanban card

the final assembly operation. This schedule is based on market demand. The objective of the kanban logistical system then becomes one of trying to keep the product flowing as smoothly as possible with a minimum of disruptions. In the JIT process of continuous improvement, the Japanese try to remove inventory and excess capacity from the operation. When these actions result in a disruption of the product flow, the workers and managers quickly work to identify the cause of the disruption and attempt to solve the problem by improving the system. If a quick solution to the problem is not found, then the previously removed inventory and workers are quickly reinstated.

Principle 4: The level of utilization of a non-bottleneck resource is controlled by other constraints within the system.

Principle 5: Resources must be utilized, not simply activated.

Synchronous manufacturing principles 4 and 5 are closely followed in JIT systems. It has already been mentioned that the work stations must receive permission to work (e.g., in the form of kanban cards) or no work will be done. This is total adherence to principle 5. The Japanese realize that activation without need, simply to keep workers busy, only results in excess inventory, which in turn disrupts the smooth product flow. Since non-bottleneck resources have excess capacity, they cannot spend all of their available time processing material. The actual level of utilization for non-bottlenecks is determined by the final assembly schedule, by the way the kanban cards arrive at the work stations, and, ultimately, by the bottlenecks in the system.

Principle 6: A transfer batch may not, and many times should not, be equal to the process batch.

Principle 7: A process batch should be variable both along its route and over time.

The JIT approach as exemplified by the Toyota kanban system fully recognizes the importance of synchronous manufacturing principles 6 and 7. In the kanban environment, the transfer batch is one standard container of material since materials are moved one container at a time. Conversely, the process batch may be any size, from one container to a very large number of containers. If the work station processes only one material and always to the same specifications, then the process batch is practically infinite. At the other extreme, since the order in which the production kanbans arrive at the work station determines the order in which the containers of material are processed, the process batch could easily be as small as one container. In general, principle 6 is valid in the kanban system. It is also easy to see that at a particular work station, given the random arrival of production kanban cards, how the process batch changes from one point in time to another. Furthermore, the arrival of the production kanbans at other work stations in the line determines the process batch sizes for those stations. Thus, in a kanban system, the process batch is variable and continually changes.

Limitations of the JIT Approach   The Japanese have used the just-in-time philosophy with enormous success in repetitive manufacturing industries. There are many elements of JIT that are helpful in implementing synchronous manufacturing, such as preventive maintenance, statistical process control, multiskilled workers, and setup reductions. But the JIT approach and JIT logistical systems, such as kanban, are not without their limitations. It is crucial that these limitations be thoroughly understood in order that a better and more general approach to synchronous manufacturing may be developed. JIT systems have at least four significant limitations that should be mentioned. First, the number of processes to which JIT logistical systems such as kanban may be successfully applied is limited. Second, the effects of disruptions to the product flow under a kanban system can be disastrous to current through-put. Third, the implementation period required for JIT/kanban systems is often lengthy and difficult. Fourth, the process of continuous improvement inherent in the JIT approach is systemwide and therefore does not focus on the critical constraints, where the greatest gain is possible. Each of these limitations is now explained in detail.

Limited Applicability   The implementation of JIT logistical systems is essentially restricted to repetitive industries. It is not appropriate to implement a JIT logistical system in either continuous-flow (process industry) environments or in plants that do not produce relatively large quantities of standardized products.

In process industries, finished goods are not produced in discrete units. Process industries are typically highly capital intensive with the system specifically designed to produce large batches of a product. For reasons of processing efficiency, this type of operation simply cannot be switched continuously back and forth from one product to another. In addition, the amount of work-in-process inventory in the system is generally set by the constraints and capacities of the system.

Industries that do not produce a large volume of fairly standardized products (i.e., job shops) will have trouble attaining the JIT objectives of small lot sizes and reduced levels of inventory. The basic explanation of why job shop operations have difficulty implementing JIT logistical systems is simple. In order to keep a smooth product flow throughout the process, safeguards must be provided at every work station in the form of inventory buffers. Kanban logistical systems require that at least two standard-sized containers of every work-in-process material, component part, or subassembly be carried in inventory at every step of the process for every unique product that is produced. This is necessary because at least one container of each type of work-in-process material must be carried at both the inbound stock area of the producing work station and the outbound stock area of the feeding work station. Thus, in the typical make-to-order job shop environment, the minimum amount of inventory required to make the system work would be immense. Therefore, most job shop processes are clearly not feasible candidates for kanban logistical systems.

It is possible to derive several rules of thumb to determine the potential applicability of a production process for a kanban logistical system. One simple rule is that a standardized container of parts should normally be consumed by the next step of the process on the same day that it is produced. If this condition is not met, then the level of inventory required to support the JIT system would likely be unnecessarily large.

Effect of Disruptions to the Flow   Remember that in a JIT/kanban system, work stations do not work independently and at their own pace. Each station is a link in a logistical chain. Whenever one work station experiences a disruption significant enough to cause a work stoppage, such as unavailability of required materials, poor-quality materials, or a machine malfunction or breakdown, the entire logistical system product flow is in jeopardy. The downstream stations will cease receiving inbound work-in-process materials and will not be able to produce. The upstream feeding stations will no longer be receiving signals to replenish materials and will also stop working since their outbound stock buffers will be full. The line comes to a halt, and throughput is lost until the disruption is corrected. The situation is similar to the assembly line's vulnerability to disruptions. The major difference is that the kanban system has the ability to (and typically does) carry more buffer inventory in the inbound and outbound stock areas than assembly lines using conveyors. However, the more inventory carried for protection, the less synchronized and less competitive the operation becomes.

Implementation Requirements   In order to implement a JIT logistical system successfully, a number of significant changes must occur in the manufacturing environment and management culture. Experience indicates that the length of time necessary to implement a working JIT system is generally several years. This is a very trying time, since the implementation process is very disruptive to normal operations, and requires both financial staying power and great patience on the part of management.

JIT is a very different philosophy from that found in the traditional manufacturing environment, and it requires increased reliance on a more responsible, better trained, and better educated work force. Ultimately, in order to implement a JIT logistical system successfully, quality problems must be virtually eliminated, setup times must be drastically cut throughout the operation, and other sources of production fluctuations must be significantly reduced across the board. Most of these problems are quite difficult to solve.

Unfocused Process of Improvement   There are two significant additional weaknesses in the JIT approach. These two limitations prevent the JIT logistical system from being able to focus attention effectively on those critical constraints of the process that limit overall productivity.

One weakness of the JIT approach is the inability to identify systematically the critical, capacity constraint resources in the operation in advance. The Japanese approach to attacking waste and supporting the process of continuous improvement within the plant is essentially unfocused.

The basic approach of improving the process is to wait until a problem occurs and disrupts the system. The Japanese are actually pleased when their planned actions to reduce inventory or workers from the system result in a disruption to the flow. This presents them with the opportunity to improve the operation of the system by taking corrective action on the resource that caused the disruption. Unfortunately, it is unknown whether the problem causing the disruption to the product flow has occurred at a true bottleneck resource. In all likelihood, the disruption may simply be the result of a statistical fluctuation at a non-bottleneck resource, or it may be a temporary schedule-induced problem. Thus, there is no guarantee that limited labor and capital resources are being used most efficiently to increase throughput and reduce inventory and operating expense. Managers in a JIT environment are unable to apply synchronous manufacturing principles 2 and 3 effectively.

Meanwhile, because of the disruption, the production process has been damaged (loss of throughput). Henry Ford recognized the importance of protecting throughput when he said "It is better to avoid difficulties than to overcome them. . . ." [10, p. 138] Remembering the river analogy, the JIT approach is to crash into the rocks and, once the rocks are discovered, try to remove them. A better approach, which is developed in the next chapter, is to detect the rocks in advance and steer around them until they can be removed.

A related aspect of this weakness is the JIT philosophy of improving the process everywhere in the system. According to synchronous manufacturing principles 2 and 3, the return on improvements at a bottleneck resource is enormous. But the return on improvements made at non-bottlenecks is marginal at best and often inconsequential. In other words, whether across-the-board improvement activities have any impact on the organizational goal of making money is unknown.

Another weakness of JIT logistical systems is the inability to preplan the production schedule for any resource in the process except final assembly. The assembly schedule does not consider the resulting loads at the bottleneck work stations. Furthermore, lead times are not taken into account except at final assembly. The JIT logistical system must continuously resort to using trial and error to deal with the changing realities and complex interactions within the system. Consequently, the production schedules may not effectively utilize the bottleneck resources and capitalize on synchronous manufacturing principles 2, 3, and 6. And since bottlenecks determine the throughput for the entire system, the resulting throughput may be less than optimum.

In the next section, the general philosophy of synchronous manufacturing is presented. In Chapter 6, a logistical system that effectively eliminates all of the limitations of assembly line and JIT systems is developed. This logistical system is consistent with, and fully utilizes, all of the principles of synchronous manufacturing.

THE PHILOSOPHY OF
SYNCHRONOUS MANUFACTURING

Not every manufacturing organization can adopt the principles of assembly lines or just-in-time systems to their operations. Nor would they necessarily want to, in light of some of the limitations identified in this chapter. However, all firms can and should adopt the philosophy and principles of synchronous manufacturing. The assembly line and JIT management systems inherently recognize and partially apply some of these principles in a limited number of processing environments. But it is time to recognize that much more is possible. If manufacturing organizations are to achieve their competitive potential, a different approach is necessary.

As principles, logistical systems, techniques, and appropriate managerial behaviors are developed and applied throughout the remainder of this text, it is important to keep in mind that synchronous manufacturing is not simply a productivity program or material control system. It is much more than that. [2, p. 108]

> *Synchronous manufacturing* is an all-encompassing manufacturing management philosophy that includes a consistent set of principles, procedures, and techniques where every action is evaluated in terms of the common global goal of the organization.

Implementing the Synchronous Manufacturing Philosophy

Synchronous manufacturing can produce rapid improvements in most manufacturing environments because it provides the means to identify and focus on the common goal of the firm. Every program, every decision, and every activity is evaluated in terms of whether it contributes to the successful accomplishment of the common global goal. In order to implement the synchronous manufacturing philosophy successfully, three major elements are necessary:

1. Define the common goal in terms that are understandable and meaningful to everyone in the organization.

   Synchronous manufacturing introduces a new set of operational measures that define the common goal of making money in terms that are easily identifiable and meaningful to everyone in the organization. These operational measures are the previously developed concepts of throughput (T), inventory (I), and operating expense (OE). These three measures are universally acceptable because they are intrinsic to every manufacturing environment, and they measure global performance for the entire system. By shifting attention from the traditional localized cost and performance measures to these global measures, synchronous manufacturing makes it possible for each individual to relate to the common goal.

2. Develop the causal relationship between individual actions and the common global goal.

   Synchronous manufacturing establishes a set of universal principles that enable us to use effectively the global measures of T, I, and OE to relate specific manufacturing actions to the enhancement of the common global goal. Some of these principles have already been presented and can be used to evaluate alternative marketing and production strategies in terms of their impact on overall T, I, and OE. They can also be utilized to establish the appropriate material and production control systems that are necessary to improve the performance of the firm with respect to the T, I, and OE measures.

3. Manage the various actions so as to achieve the greatest benefit.

   Synchronous manufacturing provides the foundation that enables managers to develop and implement appropriate material and production control systems within complex manufacturing environments. This foundation includes procedures for identifying and effectively managing all the constraints of the system and for establishing a program of focused ongoing improvement. To analyze an operation effectively, all organizational constraints—market, capacity, material, logistical, managerial, and behavioral—must be identified. Then with a properly focused analysis, policies and procedures designed to provide for the synchronous operation of the plant can be developed. The current policies and procedures most likely to be in conflict with improving the organizational competitiveness also can be identified and, ultimately, either eliminated or modified. The synchronous manufacturing philosophy enables managers to zero in on those areas of the operation that offer the greatest potential for global improvement. This process of focused improvement becomes a vital part of a program of continuous improvement throughout the entire organization.

Synchronous Manufacturing as an Umbrella

Synchronous manufacturing is a management philosophy that provides the umbrella under which all other programs, such as automation and quality improvement, should be implemented. It helps management to focus on the critical issues that impact the common global goal. Proper perspective and focus are the keys. Once the critical issues and questions have been identified, the synchronous manufacturing philosophy, principles, and techniques can be applied by the appropriate functional group in the organization. The synchronous manufacturing approach provides the framework for correctly addressing key questions such as:

- Which programs are appropriate for achieving the firm's competitive goals?
- What areas of the plant offer the greatest opportunity for improvement?
- Where should programs/changes be implemented first?
- What is the impact of the proposed change on the plant as a whole?

The analysis in previous chapters has demonstrated that the traditional cost-based system does not provide very reliable answers to these questions. Furthermore, our experience supports the conclusions of that analysis. For example, our experience reminds us of numerous worthwhile productivity and quality improvement programs that cannot be justified by a traditional cost/benefit analysis. Many of the programs that executives intuitively feel must be undertaken to remain competitive—such as "Quality Is Job 1"—have to be justified outside the framework of the traditional cost system.

Moreover, the benefits of these programs do not appear as quickly as expected or in the magnitude expected because the focus of the implementation is often incorrect. Without the kind of focus and impetus that synchronous manufacturing is able to provide, most programs will continue to be less successful than expected and less successful than they can be. [2, pp. 112–114]

SUMMARY

Synchronous manufacturing is not a totally new philosophy. In fact, some of the basic principles of synchronous manufacturing were first applied on a wide scale by Henry Ford 70 years ago. As the decades passed, many of the principles first practiced by Ford slipped into virtual obscurity until revived under the guise of a new just-in-time philosophy.

In this chapter, we have seen why assembly lines and JIT systems work and how they incorporate some of the basic principles of synchronous manufacturing. But we have also identified their limitations. Foremost among these limitations is the fact that assembly lines and JIT systems are based upon logistical systems that cannot be effectively applied to all types of manufacturing environments.

Synchronous manufacturing is an all-encompassing manufacturing management philosophy that includes a consistent set of principles, procedures, and techniques where every action is evaluated in terms of the common global goal of the organization. The basic principles of synchronous manufacturing can be applied to any organization.

QUESTIONS

1. Why can we say that Henry Ford was the first JIT manufacturer?
2. How do the Japanese practice the concept of respect for workers?
3. "The decline of U.S. manufacturing coincides with the rise of cost accounting." Discuss this statement.
4. In the river analogy, the physical factors that cause us to carry inventory (such as absenteeism) may be compared to boulders in the river. The use of very large batch sizes to save on setups also causes excess inventories and can be thought of as a managerial boulder. List some additional managerial boulders.
5. In the traditional EOQ diagram, the carrying cost is assumed to be linear. How would the carrying cost curve be modified if the true competitive cost of inventory is considered?
6. In the traditional EOQ diagram, how should the setup cost curve be modified for (a) bottlenecks and (b) non-bottlenecks? (Hint: Use synchronous manufacturing principles 2 and 3.)

7. Discuss how assembly lines and kanban systems do not explicitly adhere to synchronous manufacturing principles 2 and 3.
8. What kind of manufacturing environments are suitable for assembly line processes? What kind of manufacturing environments cannot use assembly lines?
9. What kind of manufacturing environments are suitable for kanban-type JIT systems? What kind of manufacturing environments cannot effectively use this type of system?

PROBLEMS

1. Consider an operation where a single product is processed sequentially through five resources (R1, R2, R3, R4, and R5). The processing time per unit for resources R1 through R5 is 10, 15, 8, 20, and 10 minutes, respectively. Construct a time chart showing the flow of material through all five resources for the following batch sizes: process batch of 100 units coupled with a transfer batch of (a) 100 units, (b) 50 units, (c) 20 units. For each different transfer batch size, calculate the resulting production lead time and the in-process inventory in the system.
2. Consider an operation where a single product is processed sequentially through four resources (R1, R2, R3, and R4). There is only one machine at each resource except for resource R3, which has two machines. The machines at R1, R2, and R4 have processing times of 10, 15, and 15 minutes per unit, respectively. The two machines at resource R3 can each process units at a rate of 25 minutes per unit. Construct a time chart showing the flow of material with a process batch and transfer batch both equal to (a) 100 units, (b) 50 units, (c) 20 units.

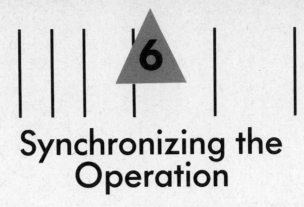

Synchronizing the Operation

PLANNING FOR A SYNCHRONIZED FLOW

In order for a manufacturing plant to evolve into a synchronous operation, the objectives of the firm must be represented in meaningful terms that can be implemented. Once the objectives have been established, a manageable approach for achieving the objectives must be selected.

The Objective

From the manufacturing manager's perspective, the overall objective is to manage the workers, machines, and materials in such a way that the demands of the marketplace are met while keeping the total costs of the system to a minimum. There is little disagreement on the objective. But there are significant differences of opinion on how best to achieve the objective.

In most plants the assumption is made that the way to keep total costs to a minimum is to minimize the production cost of each individual product. The methods by which product costs are calculated creates a situation in which the demands for the reduction of product costs (e.g., larger batches and fewer setups) are in conflict with the demands for a fast, smooth material flow (e.g., smaller batches and more setups). Unfortunately, since product cost has been the dominant measure used by manufacturing managers, efficiency has almost always taken precedent over product flow.

As discussed in Chapter 2, the assumption that reducing individual product costs will reduce total costs is not valid because of the complex interactions that exist in manufacturing operations. Thus, the procedures used to evaluate and manage manufacturing operations must be changed. At the same time, the approach to managing the product flow has to be redefined. One expedient way to do this is to use the operational measures of throughput, inventory, and operating expense. In most cases, the throughput expectations are determined before the manufacturing and production managers get involved. As a result, the goal of managing the production flow can generally be stated as *meeting throughput expectations while efficiently managing inventory and operating expense.* [3, p. 5]

Inventory and Operating Expense Tradeoffs

The marketplace is the ultimate constraint on the rate of throughput for the firm. But once the target rate of throughput has been determined, the proper tradeoffs between inventory and operating expense must be made. It must be recognized that there is no simultaneous minimum of both I and OE. It is always possible to increase OE to achieve additional reductions in I, and it is likewise possible to increase I to achieve additional reductions in OE. For example, the ideal production flow from the perspective of minimizing inventory would be to match production exactly with market demand and use batches of size one. The inventory in such a system would be extremely low. However, the expense of operating the facility would be prohibitive. Clearly, the most efficient way to operate a production facility is not to operate in a zero-inventory mode. A similar argument can be developed which demonstrates that a manufacturing environment where operating expense has been "cut to the bone" is also not optimum.

The lesson to be learned is that there are a large number of levels of I and OE which can support the desired level of throughput for a firm. Tradeoffs between I and OE will always be possible. But there is no magic formula that automatically identifies the optimum levels of I and OE in a given environment. In fact, since the system constantly changes and can always be improved, such optimum levels do not exist.

Remember that I and OE both contribute to cost, and the level of work-in-process inventory is a key determinant of manufacturing lead time and competitiveness in the marketplace. The ideal way to ensure improvement of the system is to increase throughput while simultaneously decreasing inventory and operating expense. The key to achieving this improvement lies in developing a management control system that successfully applies the principles of synchronous manufacturing.

Choosing a Manageable Approach

Planning the flow of material through a manufacturing operation in order to achieve a high level of customer service while controlling cost is a critical managerial function. This function is complicated by the fact that the number of different resources and products tend to be large in most plants. The information required for making good decisions is not always accurate or even available. In many cases, key information is extremely dynamic and difficult to track and update (e.g., work-in-process inventory). In addition, the system is susceptible to fluctuations such as machine breakdowns, absenteeism, and yield problems. The task of developing and managing a solution to this complex problem is very difficult. [3, pp. 10–11]

There are a large number of possible management control systems that can be developed and used to manage the product flow. There is also an interesting general relationship that exists between the degree of sophistication and the manageability of the management control system. This relationship is graphically demonstrated in Figure 6.1. As indicated in the figure, there are two extreme approaches to developing a management control system, both of which result in unmanageable situations.

One approach is to base the development of the management plan on a complex and sophisticated system that tries to account for all of the varied and complex interactions of the manufacturing environment. Such an approach results in the following developments:

1. A detailed plan is made that can be executed (at least in theory) at each work center.
2. Accumulation and maintenance of large amounts of data is necessary.

FIGURE 6.1   THE RELATIONSHIP BETWEEN THE LEVEL OF SOPHISTICATION AND THE MANAGEABILITY OF MANAGEMENT CONTROL SYSTEMS

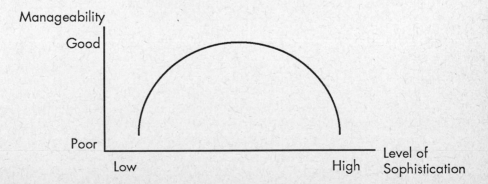

3. An enormous amount of managerial time is required to ensure data integrity and to react to the nonmodeled dynamics of the real world.

Since this approach relies on continually accurate data from all aspects of the operation, it is virtually unmanageable. The instability and unmanageability of such a tightly constructed plan means that the actual performance will be significantly and unpredictably different from the plan. The system is fatally flawed.

A second approach is to base the development of the plan on a simplistic system and use managerial expertise to compensate for the shortcomings of the logic. Such an approach results in the following developments:

1. The resulting plan is one that cannot be executed in detail, even in theory.
2. Because of the inherent shortcomings of the plan, managerial time is constantly spent making adjustments to the plan.
3. Since managerial adjustments to the plan are performed at each step and in an uncoordinated fashion, total plant performance slips. Managerial time and effort is also increasingly diverted to reacting to crisis situations.

Both of these approaches result in situations that are difficult to manage. But more important, the performance of the plant is likely to be unsatisfactory. We now consider a logistical system that avoids the problems of these two approaches.

THE DRUM-BUFFER-ROPE PHILOSOPHY OF SYNCHRONIZATION

In order for a manufacturing plant to achieve the benefits of a synchronous operation, a logistical control system that is manageable and results in predictable performance is required. The generalized system described in the remainder of this chapter, referred to as the drum-buffer-rope (DBR) system, satisfies these requirements.

Overview of the Drum-Buffer-Rope Approach

There are two fundamental issues that must be addressed when trying to develop a synchronized product flow. The first issue is the ability of the plant to execute the planned product flow over a given period of time. The second issue has to do with the impact of ever-present deviations on the planned product flow. The DBR approach explicitly considers both issues.

Defining the DBR Elements   From the perspective of developing good, workable production plans, the critical constraints in a manufacturing plant are market demand, capacity, and material limitations. When deriving the basic production plan, these constraints must be specifically considered. First, the proposed production plan quantities should not exceed projected market demand. Second, there must be a sufficient supply of materials available to support the production plan. Finally, the proposed product flow required to support the production plan must not overload the processing capabilities of the resources. This process results in the identification of a general level of production that both the market and the plant can support. Such a plan is necessary in order to avoid missed deliveries, production mismatches of components, and general confusion on the shop floor.

Once a preliminary production plan has been established, the next step is to develop detailed schedules for the capacity constraint resources (CCRs) in the plant. The CCR schedules can be used to finalize the production plan. The resulting modified production plan, called the master production schedule (MPS), becomes the basis for scheduling actual production on the shop floor and for making promises to customers. The process used to establish the MPS is referred to as the *drum*.

Of course, there are always disruptions and variances in any manufacturing process. The fact that there are deviations indicates that the actual material flow through the system will not be exactly the same as the planned flow. The only way that promises to customers can be kept is through the use of a protective cushion or *buffer* for the planned product flow.

In the case where a given product is composed of just one part, with just one simple linear routing, the following equation indicates how a time buffer is viewed for planning purposes:

Planned lead time = Sum of process and setup times + Time buffer

However, if the product is assembled from two or more component parts, then the situation is somewhat more complex. In such a case, simply consider that the planned lead time for any part can be calculated according to the planned lead time formula just stated. (The routing for an individual part may not include a time buffer.) The planned lead time for the product is then calculated as the longest of the planned lead times for its component parts plus any additional time buffer that may be provided at assembly.

In effect, the time buffer increases the planned lead time from the absolute minimum required to machine/process the products by an amount sufficient to accommodate the disruptions that are likely to occur in the manufacturing process.

The MPS, which is based on the processing capabilities and requirements of the CCRs, determines the pace and the sequence of production for the entire plant. The MPS itself is analogous to a *drum beat* used to control the pace of a march. Once the MPS is determined, the product flow for the entire plant can be planned. The product structure, process sheets, buffers, and production policies are used to determine the quantity and timing of planned production at each resource. Thus, the planned production schedule at each resource is tied to the pace set by the MPS or drum beat. The schedules of all non-CCRs will be synchronized to fully support the MPS, which in turn is structured to reflect and support the schedules of the CCRs. This part of the planning strategy is analogous to tying mountain climbers or marching soldiers together with ropes to ensure that all members proceed together at a relatively synchronized pace.

In actuality, the *rope* is a methodology to ensure the required synchronization at all non-CCRs without having to actively control each individual resource. One function of the rope is to generate the timely release of just the right materials into the system at just the right time. If the drum beat, the buffers, and the ropes are all properly determined and managed, then the desired throughput of the system will be protected, and the resulting levels of inventory and operating expense for the plant will be relatively small.

The DBR Strategy   The essence of the DBR philosophy can be briefly summarized as follows:

1. Develop the MPS so that it is consistent with the constraints of the system. (Drum)
2. Protect the throughput of the system from the inevitable minor fluctuations through the use of time buffers at a relatively few critical points in the system. (Buffer)
3. Tie the production at each resource to the drum beat. (Rope)

This sounds remarkably similar to the standard prescription for obtaining valid production schedules using MRP and MRP II systems. The concept that the MPS must be different from market demand has long been an integral part of MRP knowledge. However, it has not been successfully integrated into the MRP scheduling process for at least three reasons. First, no systematic procedure for developing a valid MPS is currently available. Given the difficulties that firms experience in gathering and verifying the vast amounts of required manufacturing data, managers cannot rely completely on computerized data files. Second, detailed techniques used with MRP systems— such as batch sizing, safety queues, transit times—are designed to permit local optimization at each work center and are not designed to promote a fast and smooth flow of material through the entire plant. Third, there has been little recognition of the basic conflict that exists between the requirements for success of MRP systems (such as schedule adherence) and the prevailing infrastructure of management policies and performance evaluation procedures.

The DBR approach, which is based on the concepts of synchronous manufacturing, differs significantly from other planning and control systems in three fundamental ways:

1. The DBR system begins with an analysis of the requirements for achieving a smooth and fast flow of material through the plant. The detailed techniques within the DBR system used to determine batch sizes and schedule priorities are developed to support global objectives, not local ones.
2. Conflicts with the infrastructure are explicitly recognized and resolved.
3. Systematic procedures for managing complex factories with their data problems and unpredictable disruptions have been developed.

Now that the overall logic for establishing a good production plan has been presented, the focus for the rest of this chapter shifts to the technical concepts of the DBR approach, especially as they relate to items 1 and 3 just presented. In the remainder of this chapter, the individual elements of the DBR approach—the drum, the buffer, and the rope—are discussed, and systematic procedures for their management are developed.

The Time Buffers

The basic concept of time buffers has already been introduced in the DBR overview section. In this section, a full explanation of the need for time buffers, the placement of time buffers, and the appropriate size for time buffers is presented.

Accounting for Disruptions   Disruptions reduce the available productive capacity of a resource. But it must be clearly understood that not all disruptions have the same impact on the product flow. A machine that is down for 2 hours represents a lost capacity of 2 hours at that resource. Disruptions at non-CCRs can impact the timing of the product flow, but do not directly affect the quantity of product produced. However, if the disruption occurs at a CCR/bottleneck resource, then this can affect the quantity and timing of the product flow and adversely impact the performance of the entire plant!

Another aspect of disruptions is that they are unpredictable. There is no way to know in advance which tool will break or which machine will malfunction and produce defective material. Thus, given the random nature of disruptions, how can reliable and stable material flow plans be developed?

The answer lies in the realization that our goal is not to make the actual product flow exactly the same as the planned product flow, but to make the actual product flow sufficient to satisfy the market demand. That is, the planned product flow should be developed in such a manner that, despite the occurrence of disruptions, the actual product flow will satisfy demand.

Developing Time Buffers—A Simple Case   Disruptions at non-CCRs can be dealt with quite easily. Since the only effect of such a disruption is to delay the timing of the material flow, it can be handled by allowing some extra time in the planned product flow.

Figure 6.2 illustrates a sequence of five operations required to produce a particular product. The illustration also shows a customer order for 10 units of the product, which will require a total of 40 hours to complete. Assume that each of the five operations is performed by a different resource, none of which is a CCR. Further suppose that the first operation is actually scheduled to begin processing 40 hours before the order is due for shipment. In this case, if there are any disruptions in the process, the shipment will be delayed.

FIGURE 6.2   SIMPLE FIVE-STEP LINEAR PROCESS FOR A 10 UNIT BATCH OF PRODUCT A

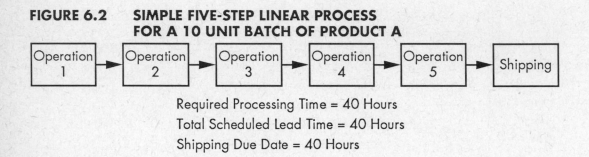

Required Processing Time = 40 Hours
Total Scheduled Lead Time = 40 Hours
Shipping Due Date = 40 Hours

Now suppose that a safety cushion of 20 hours is introduced into the system. This means that the first operation begins processing 60 hours before the order is scheduled to be shipped. Of course, if there are no disruptions, then the order will be finished 20 hours before it is required. It will simply be placed in inventory for a short period of time. One might presume that the order can be shipped on time as long as the disruptions total less than 20 hours. We will demonstrate, however, that this is not necessarily true. We will also demonstrate that disruptions affecting the product flow can be effectively handled by the proper placement of protective buffers.

It is important to understand that we are not suggesting that the process be protected by developing excess stocks of inventory. Instead, the process is protected by allowing for excess time in the schedule (*a time buffer*), which may in turn cause some additional inventory in the system.

The Placement of the Time Buffers   One way of providing the 20 hour time buffer in the previous example is to provide five time buffers of 4 hours each. This configuration is illustrated in Figure 6.3. The first four time buffers are placed between operations, and the last buffer is placed after the final operation and before shipment of the order. This means that each of the

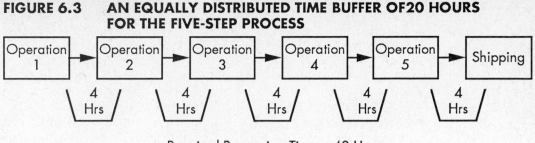

Required Processing Time = 40 Hours
Total Scheduled Lead Time = 60 Hours

first four operations is scheduled to complete its work 4 hours before the next operation is scheduled to start. The last operation is scheduled to complete its work 4 hours before the order is needed for shipping.

This buffering system clearly affords some protection for each operation as well as for shipping. For example, if operation 1 experiences a disruption of 4 hours or less, operation 2 is not affected. The same is true for any of the last three operations, as well as for shipping. Shipping and operations 2 through 5 are protected from short-duration disruptions at earlier operations by the 4 hour time buffers.

In this buffer configuration, the system as a whole can often, but not always, be protected from more disruptions than any single operation. Consider the effect of a significant disruption. Suppose, for example, that a 12 hour disruption occurs at operation 1. Even then, there is no immediate danger to on-time shipment of the order. The start of operation 2 will be delayed by 8 hours. If there are no further disruptions or delays at operation 2, then the start of operation 3 will be delayed by 4 hours. If there are no delays at operation 3, then operation 3 will finish its work 4 hours behind schedule. Since operation 4 has a 4 hour time buffer, it can start its work on schedule. If this is the only disruption in the process, then the order will be shipped on time.

But what is the resulting impact on the shipment if a 12 hour disruption hits operation 5? Since there is only a 4 hour buffer between operation 5 and shipping, the order will be late by 8 hours.

This philosophy of distributing the time buffers throughout the process has the following features:

1. Each operation is buffered or protected to a certain extent.
2. The system has, in general, more buffer than any single operation.
3. The amount of buffer available depends on the location of the disruption in the process. The further along in the process the disruption occurs, the less protection there is.

It should be noticed that this buffering approach is somewhat similar in effect to a kanban system.

It is vitally important to remember that it is not necessary to protect every operation in the plant. It is only necessary to protect the overall product flow for the system and, in particular, shipments. If our attention is properly focused, then the approach to the placement of buffers can be improved.

Instead of a system where the buffers are distributed throughout the process, consider the case in which the entire buffer is provided at the end of the process. That is, all 20 hours of the time buffer are placed after operation 5, before shipping. This configuration is illustrated in Figure 6.4. This scenario means that operations 1, 2, 3, and 4 are planned to be completed at the exact moment that material is required by the next operation. Operation 5 is planned to be completed 20 hours before the product is due to be shipped.

FIGURE 6.4   A 20 HOUR TIME BUFFER PLACED ENTIRELY BEFORE SHIPPING

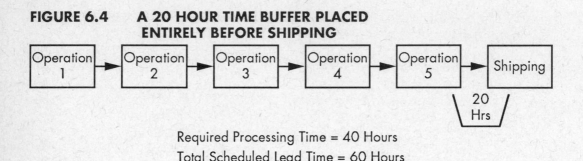

Required Processing Time = 40 Hours
Total Scheduled Lead Time = 60 Hours

Now consider the effect of a 12 hour disruption in this new configuration where all of the buffer is after operation 5. A 12 hour disruption at operation 1 will cause a delay of 12 hours at operation 2. If there are no further delays, then operations 3, 4, and 5 will each be delayed by 12 hours. Since operation 5 is planned to be completed 20 hours before the due date, the effect of the disruption is that operation 5 will complete the order 8 hours before it is due to be shipped. Significantly, the exact same argument applies to any disruption, no matter where it occurs. All operations that are downstream from the disruption will be delayed, yet the total system is protected by the full buffer. Any disruption or combination of disruptions in the system that do not exceed 20 hours will not affect the timely shipment of the order.

This configuration, where the entire buffer is placed before shipping, has the following characteristics:

1. The individual operations are not protected from disruptions. Since disruptions will invariably occur, most operations will often be behind the plan.

2. The total system has the benefit of the full buffer, no matter where the disruption occurs.
3. The existence of one well-defined buffer, precisely located, serves to focus attention on it. The reason for its existence is constantly in mind. Significant problems with getting material to the buffer in a timely manner are highly visible.

Given these circumstances, it is evident that having the buffer concentrated at the end of the process is better than having it distributed throughout the process. However, the foregoing illustration was somewhat oversimplified, since there were no CCRs or bottleneck resources in the process. The only real constraint on the process was the shipping due date. But in many manufacturing plants, placing time buffers only at the end of the process is not sufficient.

Protecting the CCRs   Disruptions that affect both the quantity and timing of the product flow, such as those that occur at true bottlenecks, cannot be solved in the simple fashion we have just described. They must be dealt with directly. The problem must be resolved as quickly as possible, and additional capacity must be made available to make up for lost time.

A complicating factor in the interacting environment of manufacturing is that disruptions can affect not just the operation in question but other operations as well. Unless special precautions are taken, a delay at a particular operation will delay all succeeding operations. Similarly, a piece of material scrapped at a particular station will require that all preceding stations make up this lost material. Thus, disruptions spread. This means that a simple disruption at a non-CCR can become a serious disruption at a bottleneck or CCR. The product flow must be protected from such dangerous spreading. The actual material flow through the CCR should be the same as the planned product flow. Otherwise, the product flow of the system may be adversely affected.

To protect the product flow through the system, the product flow through the CCRs must be protected from disruptions at other stations. This can be accomplished by allowing some extra time in the planned product flow for material scheduled to arrive at the CCR operation. That is, time buffers must be provided in front of CCR operations.

It is evident that in order to maintain our ability to process and ship products on time and in the quantity required, time buffers will generally be required in at least two places. The time buffers are needed (1) at the end of the process, before shipping; and (2) in front of the CCRs in the process.

Continuing with our previous illustration, now suppose that operation 3 is a CCR. To protect the throughput of the process, time buffers are established in front of operation 3 and before shipping. Figure 6.5 illustrates this scenario, showing a 10 hour time buffer in front of both operation 3 and shipping.

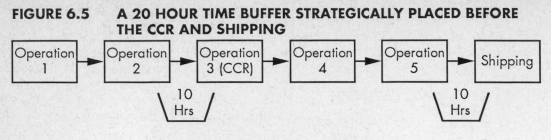

Required Processing Time = 40 Hours
Total Scheduled Lead Time = 60 Hours

When the entire production process consists of a simple linear flow, time buffers before CCRs and shipping will be sufficient to protect the throughput of the system. In more complex flows, the location of the time buffers will include other locations in addition to the two we have just identified. These locations will have to be chosen so as to buffer the product flow from potential disruptions anywhere in the system. The exact locations of these buffers are dependent upon the manufacturing process. [5, pp. 100–105]

The Size of the Time Buffers   The next key question to consider is how management should choose the size of the time buffers to be established in the system. The answer can be summarized in a very general way. If a specific time buffer is not large enough to protect the system, then the size of that buffer should be increased. For example, if a CCR is continually starved for material because of disruptions to the product flow, the size of the buffer in front of that CCR should be increased. Likewise, if the schedule of the CCR is constantly being affected by missing orders, the buffer should probably be increased. If shipments are continually late because orders are not completed on time, the size of the time buffer before shipping should be increased. These increases in the time buffer will cause orders to be released earlier. Thus, orders should arrive earlier to the buffer, resulting in fewer problems.

Conversely, consider the situation where the actual flow of material usually matches the planned flow, and the throughput is never threatened by disruptions. In this case, the time buffers may be too large. A time buffer that is too large implies excessive protection in the system. This excessive protection may make management more comfortable, but it increases WIP inventory, costs valuable production lead time, hinders market responsiveness, and reduces a firm's competitive advantage. In such cases, the time buffers should be decreased.

On one hand, management should try to protect the product flow of the CCRs and the shipping due dates. On the other hand, it is impossible to protect the system against every potential disruption, or combination of disruptions, that might possibly occur. Management should strive to identify the minimum levels for the time buffers necessary to provide adequate protection for the system. For example, suppose that most disruptions that occur at all of the operations preceding a CCR can be made up within 10 hours. In that case, a 16-hour (2 day) time buffer would probably be quite sufficient.

We have found by experience that for a firm trying to implement a DBR system, a *convenient* starting point for the size of the total time buffer is approximately one-half of the firm's current manufacturing lead time. This has two major advantages:

1. It provides sufficient buffer to make customer promise dates attainable.
2. It meets the competitive need to significantly reduce lead times.

The first point should be evident. However, the second point may require further clarification as to why including a time buffer of one-half the production lead time will reduce lead time. The basic reason is that a large majority of a product's manufacturing lead time is generally spent waiting in queues at various work stations in the system. In the DBR system, the only planned queues coincide with the placement of the time buffers, and these are very limited in number. As a result, since planned waiting time is *reduced* to about half of total lead time, the total lead time itself should be reduced.

To further illustrate the concept of a time buffer, Table 6.1 shows the planned schedule for a CCR for the coming 40 hour week (one 8 hour shift for each of 5 days). Now suppose that the CCR has been provided with a 24 hour (3 day) time buffer. This means that in the planned product flow, the CCR is expected to have 24 hours' worth of work in its queue waiting to be processed. That is, on Monday morning the expected queue at the CCR should consist of the work orders scheduled to be processed over the next 24 working hours (the orders scheduled to be processed on Monday, Tuesday, and Wednesday). According to the information provided by Table 6.1, the jobs that should be in the queue on Monday morning are work orders A, B, and C. (On Tuesday morning, the remainder of work order B, work orders C, D, E, and F are planned to be in the queue for processing.)

Disruptions are normal in all manufacturing systems, and the actual material flow will not be equal to the planned flow. Therefore, it would not be unusual if some of the work orders that are planned to be in the time buffer are missing. This will only cause a problem for the overall product flow in two cases. In the first case, suppose the queue is completely empty and the CCR is forced to shut down. This will disrupt the timing of the product flow, and, if the CCR is a bottleneck, it will also cause a loss of throughput. In the second case, suppose that some of the work orders are available, but the job with the highest priority is missing (work order A in our illustration). In this case, the CCR would normally work on the next

TABLE 6.1   A SCHEDULE OF WORK ORDERS FOR A CCR

DAY	ORDER	PROCESSING TIME REQUIRED (HOURS)
Monday	A	6
	B	2
Tuesday	B	8
Wednesday	B	5
	C	3
Thursday	D	2
	E	4
	F	2
Friday	G	3
	H	5

job in line (work order B) until it is completed. When work order A arrives at the queue, it will be processed as soon as the current job is completed. This situation is likely to cause a problem in the timing of the product flow. But it will not result in lost throughput, unless it occurs at a CCR that has dependent and lengthy setups, such as at a heat treat operation. In such cases, throwing the setup sequence out of order may cause lost capacity and a loss of throughput.

As long as scheduled work is available at the CCR, the product flow problems caused by disruptions are relatively minimal. If the highest priority job is always available, then product flow problems are nonexistent. A more detailed discussion of how to analyze and manage the time buffers effectively in order to focus improvement efforts within the operation is presented in the next chapter.

Time Buffers versus Stock Buffers   It is important to emphasize the difference between time buffers and stock buffers that are held (often in finished product form) in manufacturing operations. The key distinction between these two types of buffers is summarized as follows:

Time Buffers—Designed to protect the throughput of the system from the internal disruptions that continually occur in manufacturing environments.

Stock Buffers—Designed to improve the responsiveness of the operation to market demand. This is accomplished by holding inventories of finished or partially finished products in anticipation of future market demand. This allows for orders to be filled in less than the normal production lead time.

Time buffers may be thought of as serving the same function as shock absorbers in an automobile. These buffers absorb the various shocks or disruptions in the manufacturing system that would otherwise prevent the smooth and timely flow of materials moving through the CCRs and shipping. The size and location of the time buffers are determined so as to protect the quantity and timing of the planned throughput.

The size and location of the stock buffers are critically dependent on the manufacturing lead time. The shorter this lead time, the less is the need for stock buffers. And manufacturing lead time is, of course, the heart of these discussions. Much of the inventory that exists on the shop floor is a result of not recognizing the distinctions between these two types of buffers and their specific objectives.

The Drum

The derived drum beat (the MPS for the plant) is a function of the modifications required to bring the potential market demand in line with the capacity and material capabilities of the plant. As previously discussed, it is critically important to identify and focus on the CCRs of the system when trying to determine the appropriate drum beat. Otherwise, there will be time intervals when the required capacity at the CCR will exceed the available capacity to the extent that the CCR will be unable to maintain the planned product flow. Such an occurrence may cause the actual product flow to fall sufficiently behind schedule that promised shipping dates are jeopardized.

Identifying the CCRs   Identifying the CCRs in a complex manufacturing environment is not always an easy matter. The recommended approach is to use a diagnostic procedure similar to that used by a physician to identify diseases in patients. In this approach, the problems (symptoms) displayed by the plant (patient) are used to trace the cause (disease). To use this procedure effectively, one must know in advance the various effects that CCRs have on the synchronous operation of different types of manufacturing environments. At this point, it is too great a digression to discuss the different effects of CCRs in various manufacturing environments. A structure that is useful in diagnosing CCRs in different types of manufacturing plants is developed in Chapter 8.

Once a resource is diagnosed as a potential CCR, this does not automatically mean that the MPS should be modified to meet the needs of that resource. Modifications to the MPS impact the performance of the entire plant. Therefore, all modifications must be carefully considered by senior managers. They must recognize which resources control the product flow through their plants and also understand the consequences of leaving the resources alone or changing either their load or capacity. After this review and modification process is completed, a small number of critical resources will remain. These are the resources that largely influence the MPS for the plant. They are used to help determine the drum beat.

Setting the Drum Beat   Once the CCRs have been identified, the next step is to determine how best to modify the preliminary production plan to match the available capacity at the CCRs. The traditional approach has been a trial and error modification of the MPS until all resource loads are within their capacities. [19, pp. 165–180] The problem is that no formal procedure for this iterative approach exists. In practice, time pressures and the skill of the scheduler limit the quality of the final plan.

The DBR system uses a different approach to developing the MPS. [3, pp. 20–26] First, all resources where capacity is a problem (the CCRs) are specifically identified. The various orders that are to be processed through the CCRs are scheduled, utilizing all available capacity. Since the CCRs determine the quantity and timing of the throughput for the system, the final step is to use the scheduled product flow through the CCRs to determine what the MPS for the system should be. The drum beat for the system is now set.

A critical point in the DBR approach that still must be addressed is how best to plan the product flow at the CCRs. A key consideration in this analysis is whether or not the CCRs require setups.

If a CCR does not require setups, the priority sequence should basically be a function of the order due date and the remaining planned completion time of the order after leaving the CCR. The CCR should work in priority sequence and only in the quantities required to fill specific orders. Since the process batch should be equal to the order quantity, the only policy question in this case should be the size of the transfer batch. That is, how many units should be moved at a time? Smaller transfer batches should result in a faster flow of material and smaller inventories but may involve more material handling. Larger transfer batches require fewer material moves but normally result in a slower material flow and larger inventories.

If a CCR requires setups between batches, the situation is more complex. The critical problem becomes one of determining when to stop producing one product and start producing a different product. In other words, what is the appropriate process batch size? The answer to this question is complicated by the fact that the optimum process batch size may change over time and between products.

If the process batch is set exactly equal to the size of each individual order, the result may prove to be quite unsatisfactory. Even though the orders can be processed in exact priority sequence, this may require an excessive number of setups. It is possible to spend so much time in setting up the resource that the actual processing time is severely restricted. As a result, except for the first few orders, orders are unduly delayed; and the delay gets progressively worse for future orders.

Large process batches result in less time spent on setups; therefore, more production at the CCRs is possible. However, this approach has some drawbacks. A large process batch implies that several orders are combined into a single longer production run. Clearly, some of these orders are produced earlier than needed and out of their established priority sequence. This also means that orders for different products will have to wait longer in the queue and may even be delayed sufficiently to miss their due dates.

A simple example can be used to illustrate how the drum beat is established. Consider Figure 6.6, which provides demand and critical processing information for the production of two products, A and B. The illustration shows that resource R2 is a CCR for this process, and after processing at R2, products A and B are both processed at resources R3 and R4 before being ready for shipping. Resource R2 has a setup time of 40 minutes for both products A and B. The processing time per piece at R2 is 11 minutes for both A and B. Each resource has 480 minutes of available capacity per day. Finally, the planned product flow allows for 3 days of lead time (including a time buffer) after processing at R2 until the products are scheduled for shipping. The schedule of customer demand is 20 units of A and 20 units of B per day, as shown in Table 6.2(a).

The schedule of customer demand translates into a trial schedule for resource R2 as shown in Table 6.2(b). The table indicates that the proposed schedule calls for 520 minutes of processing time per day when only 480 minutes are available. It is apparent that R2 can only perform one setup per day and produce the required average of 40 units of total product per day. Therefore, the schedule for R2 should be modified as shown in Table 6.2(c). The resulting MPS is as shown in Table 6.2(d). The MPS is the drum

**FIGURE 6.6   SETTING THE DRUM BEAT:
BASIC PROCESS INFORMATION**

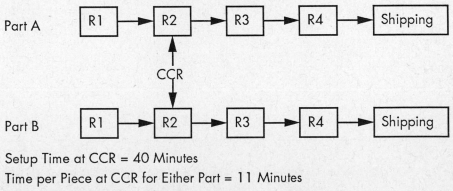

Setup Time at CCR = 40 Minutes
Time per Piece at CCR for Either Part = 11 Minutes
Lead Time from CCR to Shipping = 3 Days
Each Resource Has Available Capacity of 480 Minutes per Day

TABLE 6.2   SETTING THE DRUM BEAT:
DERIVING THE MASTER PRODUCTION SCHEDULE

(a) Customer Demand

DATE	PRODUCT	QUANTITY
7/5	A	20
7/5	B	20
7/6	A	20
7/6	B	20
7/7	A	20
7/7	B	20
7/8	A	20
7/8	B	20
.	.	.
.	.	.
.	.	.

(b) Trial Schedule at CCR (R2)

DATE	PRODUCT	QUANTITY	SETUP TIME (MINUTES)	RUN TIME (MINUTES)	TOTAL TIME (MINUTES)
7/2	A	20	40	220	260
7/2	B	20	40	220	260
7/3	A	20	40	220	260
7/3	B	20	40	220	260
7/4	A	20	40	220	260
7/4	B	20	40	220	260
7/5	A	20	40	220	260
7/5	B	20	40	220	260
.	.	.	.	.	.
.	.	.	.	.	.
.	.	.	.	.	.

beat that drives the production system. But the processing requirements at the CCR determine how the MPS should be structured in order to satisfy the existing customer demand. This illustration clearly demonstrates the significance of the system constraints in setting the drum beat.

**TABLE 6.2   SETTING THE DRUM BEAT:
DERIVING THE MASTER PRODUCTION SCHEDULE
(continued)**

(c) Modified Schedule at CCR (R2)

DATE	PRODUCT	QUANTITY	SETUP TIME (MINUTES)	RUN TIME (MINUTES)	TOTAL TIME (MINUTES)
7/1	A	40	40	440	480
7/2	B	40	40	440	480
7/3	A	40	40	440	480
7/4	B	40	40	440	480
.	.	.	.	.	.
.	.	.	.	.	.
.	.	.	.	.	.

(d) Master Production Schedule

DATE	PRODUCT	QUANTITY
7/4	A	40
7/5	B	40
7/6	A	40
7/7	B	40
.	.	.
.	.	.
.	.	.

There are three key factors that must be considered when determining the appropriate schedule for a CCR. These three factors are (1) production sequence, (2) process batch, and (3) transfer batch. The choices for these three factors determine the performance limits of the plant. Thus, we present valuable guidelines on how to modify existing production sequences, process batches, and transfer batches.

None of these three factors can be determined in complete isolation from the other two. Even though the transfer batch determination is largely independent, it is limited by the size of the process batch. However, the process batch and the product sequence are highly interrelated. A change in the process batch implies a change in the sequence, because an increase in the process batch is simply the result of combining future orders with the current order. Ideally, the sequence of production should be based on a valid system of priorities. The priorities should normally be derived from the due dates of the orders in conjunction with the expected manufacturing lead time required to complete the processing of the product after it leaves the CCR. Although modifications to this approach for determining the order priorities are sometimes warranted, it is generally sufficient.

Analyzing the Drum Beat   How should the quality of any particular production plan be evaluated? In answering this question, it should be remembered that the plan should not be evaluated from the perspective of the resource, but from the perspective of the entire plant. Thus, if the plan is good for the plant (as measured by T, I, and OE), then by definition, the plan is good for the resource. The only reason for reevaluating the plan is if it violates physical constraints at the resource—that is, the plan cannot be executed.

If the plan is not good for the plant, then it is necessary to analyze the product flow in order to identify and correct the problem. Is the sequence of production wrong? Are too many setups performed? Are too few setups performed? Are the transfer batches too large or too small?

The Schedule Performance Curve   The clue to the nature of the problem is contained in the way in which customer orders are completed and shipped. The best way to analyze the production plan is to compare the customer due dates for the various orders to the scheduled completion dates for the same orders. [3, p. 23] Such a comparison of customer due dates and scheduled completion dates is illustrated in Figure 6.7. If the scheduled completion dates are all exactly equal to their due dates, then the result will be a plot of points which yields a straight line that forms a 45 degree angle above the horizontal axis. Of course, such an exact match of due dates and scheduled completion dates will not usually occur. In reality, the resultant set of points generated by the comparison of due dates and scheduled completion dates may be summarized by any of a large number of curves. Figure 6.7 shows one such curve that oscillates above and below the 45 degree line. Those points that fall above the 45 degree line represent orders which, if completed according to schedule, will be late. Those points that fall below the 45 degree line represent orders that are scheduled to be completed early. The curve of plotted points, which compares scheduled completion dates and customer due dates, is referred to as a schedule performance curve.

The schedule performance curve has two basic functions. It can be used to illustrate the degree to which customer orders are scheduled to be finished early or late. More important, the schedule performance curve can be used to identify actions that lead to improved customer service performance.

The schedule performance curve analysis does not consider actual completion dates of the orders. It is assumed that with the established time buffers, the actual completion dates will not differ greatly from the scheduled

FIGURE 6.7   A SCHEDULE PERFORMANCE CURVE

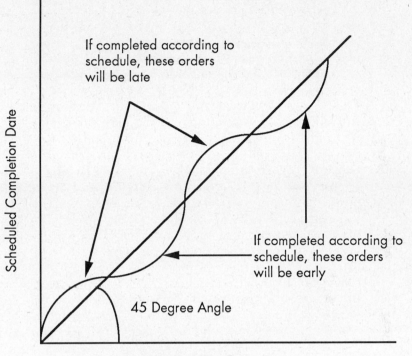

completion dates. If significant differences exist between the actual and scheduled completion dates of orders, then management's attention should be focused on reducing the disruptions that cause the actual product flow to differ from the planned product flow. (This is a primary topic of the next chapter.)

To the extent possible, the scheduled completion dates should approach but not exceed the customer due dates for the orders. To analyze effectively the proposed master production schedule, it is necessary to understand what can be learned from the schedule performance curve.

The plot of the schedule performance curve may either be random in nature or exhibit a definite pattern. If the curve exhibits total randomness, then there are major problems within the system. These problems may include a lack of proper sequencing, poor batch-sizing decisions, or significant problems in carrying out the production plan at the CCRs or at other resources later in the process. Furthermore, it will be difficult to discern what the problems actually are. However, if the schedule performance curve exhibits identifiable nonrandom patterns, some basic inferences about the plan can be made.

In general, the schedule performance curve has three features of interest. These features are: (1) the location of the curve, (2) the tilt of the curve, and (3) the oscillation of the curve.

The overall location of the schedule performance curve is influenced by the product transfer batches. A reduction in the transfer batch size of products will yield earlier completion of orders. The result of earlier completion times will be to move the schedule performance curve down and to the right. Conversely, an increase in the size of the tranfer batches will delay completion times and move the schedule performance curve up and to the left.

The tilt and oscillation of the schedule performance curve are related. It is difficult to influence one without influencing the other, because both are controlled by the process batch. If the process batch is reduced, then the oscillations will become smaller and the upward tilt of the curve will increase. This result must occur since smaller process batches imply more setups, which means that each order is processed more in line with its original priority. This in turn reduces the tendency for some orders to be early at the expense of making other orders late (thus reducing the oscillations of the curve). But the additional setups result in less total processing time, which reduces long-term throughput at the resource. Therefore, reducing the process batch at a CCR improves short-run due-date performance by sacrificing long-range throughput potential. Conversely, fewer setups at the CCR will allow for greater throughput in the long run, but only at the expense of poorer due-date performance in the short run.

Schedule Performance Curve Cases   There are a number of combinations of variations in the location, tilt, and oscillation that may be found in any given schedule performance curve. Five representative cases have been selected to demonstrate the problems that can be highlighted by the curve features. [3, pp. 24–26] These cases are illustrated by a comparison of scheduled completion dates and customer due dates for several hypothetical orders. Figures showing the general shape of the corresponding schedule performance curve (based on the hypothetical orders) are developed for each of the five cases.

> *Case 1:* The scheduled completion dates and the customer due dates for the hypothetical orders shown in Figure 6.8 indicate that every order is late by approximately the same amount of time. This results in a schedule performance curve that is entirely above and roughly parallels the 45 degree line. If all orders are consistently late, then the transfer batch sizes should be reduced. If the tranfer batches can be reduced, then the manufacturing lead time should decrease, and the curve (with its shape intact) will move closer to the desired 45 degree line. Conversely, if the orders are consistently early, then the transfer batches can be safely increased by some limited amount (if desired) without jeopardizing due dates.

FIGURE 6.8   SCHEDULE PERFORMANCE CURVE - CASE 1: SCHEDULED COMPLETION DATES LAG DUE DATES BY AN APPROXIMATELY CONSTANT AMOUNT

Order #	Customer Due Date	Scheduled Completion Date
101	7/1	7/6
109	7/8	7/12
115	7/12	7/17
120	7/19	7/21
126	7/24	7/30
132	7/27	7/30

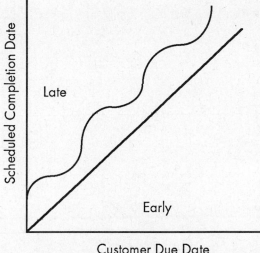

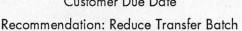

Recommendation: Reduce Transfer Batch

Case 2: Consider the situation as shown in Figure 6.9. In this case, most of the orders are completed early, but a few of the orders are late. Note also from the scheduled completion dates of the orders that it appears as if several orders are typically combined to form a single process batch. The resulting schedule performance curve has relatively large oscillations and lies mostly below the 45 degree line. The large oscillations are caused by process batches that are much larger than the order quantities. The large batches account for most of the orders being completed early. But while a large batch of one product is being processed, other orders must wait in the queue, with the possibility of being delayed past the point of meeting their due dates. The completion of orders is generally ahead of schedule and may become slightly more so as time passes. This indicates that, given the current policy decisions, a certain amount of excess capacity

FIGURE 6.9   SCHEDULE PERFORMANCE CURVE - CASE 2: LARGE PROCESS BATCHES RESULTING IN MOST ORDERS BEING EARLY WHILE SOME ORDERS ARE LATE

Order #	Customer Due Date	Scheduled Completion Date
101	7/3	7/5
109	7/8	7/5
115	7/12	7/5
120	7/19	7/20
126	7/24	7/20
132	7/27	7/20

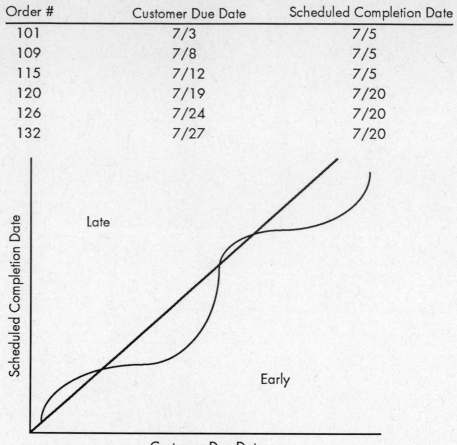

Recommendation: Reduce Process Batch

is available at the resource. To better utilize the available capacity and reduce the oscillation of the curve, the process batch should be reduced. Since the process batch can never be smaller than the transfer batch, this may also require a reduction in the transfer batch.

Case 3: The situation presented in Figure 6.10 indicates that the completion of orders consistently lags behind the due dates, and the performance is getting progressively worse. The schedule performance curve has very small oscillations, and the curve lies above and tilts away from the 45 degree line. The fact that production lags further and further behind

FIGURE 6.10   SCHEDULE PERFORMANCE CURVE - CASE 3: SMALL PROCESS BATCHES COUPLED WITH LIMITED CAPACITY CAUSE ORDERS TO BECOME PROGRESSIVELY LATE

Order #	Customer Due Date	Scheduled Completion Date
101	7/3	7/4
109	7/8	7/10
115	7/12	7/16
120	7/19	7/22
126	7/24	7/29
132	7/27	8/3

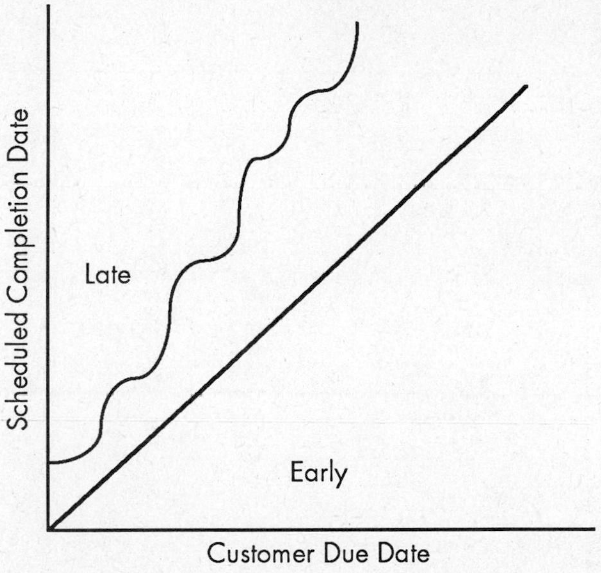

Recommendation: Increase Process Batch

the due dates indicates that the available capacity is less than what is required. The small oscillation suggests that the process batches are fairly close to the order quantities. One way to increase the productive capacity of the process in this case is to perform fewer setups at the CCRs and convert the eliminated setup time into production time. This means increasing the process batch. This will tilt the schedule performance curve back toward the 45 degree line and bring long-range throughput more in line with requirements. But it will also result in an increase in the oscillation of the curve since the process batches and order quantities are no longer roughly equal.

Case 4: In the case illustrated in Figure 6.11, the orders are scheduled to be completed before the customer due dates. Furthermore, the orders are being increasingly finished ahead of schedule. The schedule performance curve has relatively small oscillations and tilts down. In this case the process has excess capacity. Even though the process batches are roughly equal to the order quantities, production consistently runs ahead of demand. The resources in the plant are being underutilized. It may be possible to increase the throughput and profitability of the plant by capitalizing

FIGURE 6.11   SCHEDULE PERFORMANCE CURVE - CASE 4: PLANT HAS EXCESS CAPACITY AND ORDERS ARE CONSISTENTLY COMPLETED PRIOR TO THE DUE DATES

Order #	Customer Due Date	Scheduled Completion Date
101	7/3	6/29
109	7/8	7/4
115	7/12	7/7
120	7/19	7/14
126	7/24	7/18
132	7/27	7/22

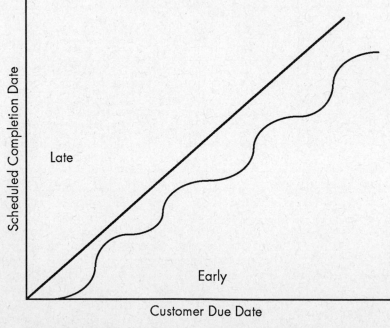

Recommendation: Increase Customer Demand

on the competitive advantage of being able to deliver with less lead time and by proper marginal costing of products. It may also be possible to better utilize the existing capacity of the resources by further reducing the size of the process batches.

Case 5: Figure 6.12 illustrates the situation where the scheduled completion dates for the orders lag the customer due dates, and the degree of lateness is getting progressively worse. The schedule performance curve has large

FIGURE 6.12   SCHEDULE PERFORMANCE CURVE - CASE 5: PLANT HAS SIGNIFICANT CAPACITY CONSTRAINT(S) AND ORDERS ARE PROGRESSIVELY COMPLETED LATE, DESPITE THE USE OF LARGE PROCESS BATCHES

Order #	Customer Due Date	Scheduled Completion Date
101	7/3	7/14
109	7/8	7/14
115	7/12	7/14
120	7/19	7/31
126	7/24	7/31
132	7/27	7/31

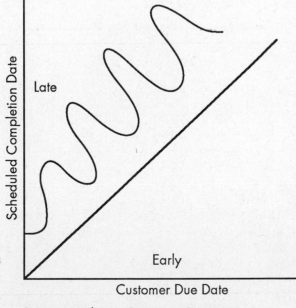

Recommendation: Increase Capacity

oscillations and tilts up and away from the 45 degree line. In this case, a real bottleneck exists. The large oscillations in the curve indicate that the production batches are already quite large relative to the order size. But even though the process batches are already large, there is still insufficient long-run capacity to meet the throughput requirements. The needed capacity cannot be generated by increasing the process batch and reducing the number of setups performed. The capacity at the bottleneck must be increased in other ways, as will be discussed in the next chapter.

These case discussions suggest that appropriate modifications to the transfer batch and process batch sizes can be determined by analyzing the schedule completion curves. Once batch size determinations have been made, the next step is to determine the impact of these choices on the expected completion dates of orders. In effect, this determines the master production schedule. In practice, this is a straightforward, but far from trivial, task.

The Rope

The *drum* provides the master production schedule that is consistent with the requirements and capabilities of the plant. The time buffers provide the insurance, at minimal cost, so that customer promises can be met with high reliability. The last link in the system is to communicate effectively throughout the plant the actions that are required to support the MPS. Every work station must be synchronized to the requirements of the MPS so that the plan may be efficiently executed. This is the function of the *rope*.

Key Considerations for Developing the Rope   Two key factors must be considered when developing the rope for a logistical system:

1. The information communicated must be meaningful. To be meaningful, the information must be relevant, valid, and not already known to the recipient. In addition, the information should provide the recipients the opportunity to use their expertise effectively.
2. Control of the schedule should not depend on the detailed management of all resources. Experience tells us that attempting to control too much detail over too wide a span of control has always been a costly and unsuccessful venture. But this should not be interpreted to mean that all details can be ignored. It simply means that in any system a relatively few critical points exist that must be carefully managed in order to control the entire system successfully. This premise has already been demonstrated in the drum process where only a few resources were relevant in developing the MPS.

A simple example is developed to illustrate the task of efficiently communicating the MPS to all work stations and effectively controlling the execution of the plan. Table 6.3 shows the MPS and the production routing files for two products, A and B. The resources required, as well as the processing and setup times at each operation, are also specified in the routings. The task is to communicate to each of the five resources what must be done to support the MPS.

Anyone who understands the rudiments of MRP or basic project scheduling is familiar with the technique for converting the MPS into a detailed schedule for each work station. The derivation of such a detailed schedule is a straight-

TABLE 6.3   A HYPOTHETICAL MPS AND ROUTING FOR TWO PRODUCTS, A AND B

(a) Master Production Schedule

DATE	PRODUCT	QUANTITY
7/1	A	20
7/1	B	20
7/2	A	20
7/3	B	20

(b) Routing File for Product A

OPERATION	RESOURCE	PROCESSING TIME PER UNIT (MINUTES)	SETUP TIME (MINUTES)
010	R1	20	30
020	R2	10	15
030	R3	15	60
040	R2	15	15
050	R4	25	10

(c) Routing File for Product B

OPERATION	RESOURCE	PROCESSING TIME PER UNIT (MINUTES)	SETUP TIME (MINUTES)
010	R1	15	30
020	R2	10	10
030	R5	30	20

forward exercise. One such MRP-type schedule is based on the assumptions that products A and B are both processed and moved in batch sizes of 20 and that there is 1 day of slack time built into the schedule at each operation. Table 6.4 contains a detailed schedule for the five resources used to produce the two batches of product A (denoted by A1 and A2) and the two batches of product B (denoted by B1 and B2) needed to satisfy the MPS. Table 6.5 shows the resulting detailed dispatch list for resource R2, sorted by scheduled start times for the four batches.

This process appears to yield a schedule that provides all the information required by each work station to support the MPS. Monitoring performance to the derived schedule provides all the information needed by managers to evaluate the execution of the plan. If each work station can meet its schedule, the plant will be able to meet its customer commitments. However, it is clear that resource R2 cannot follow the schedule as given. The operator would also question whether operation 020 for product B1 or operation 040 for product A1 should receive priority on 6/27.

The basic technique of MRP and MRP II systems is appealing. But it does not meet the requirements for effective communication and control.

TABLE 6.4   DETAILED RESOURCE SCHEDULE FOR RESOURCES R1 - R5 FOR THE TWO BATCHES OF PRODUCT A (A1 AND A2) AND TWO BATCHES OF PRODUCT B (B1 AND B2)

	PRODUCT A1		PRODUCT A2	
RESOURCE	Scheduled Start	Scheduled Stop	Scheduled Start	Scheduled Stop
R1	6/22 - 1 H 30 M	6/23 - 0 H 40 M	6/23 - 1 H 30 M	6/24 - 0 H 40 M
R2	6/24 - 0 H 40 M	6/24 - 4 H 15 M	6/25 - 0 H 40 M	6/25 - 4 H 15 M
R3	6/25 - 4 H 15 M	6/26 - 2 H 15 M	6/26 - 4 H 15 M	6/27 - 2 H 15 M
R2	6/27 - 2 H 15 M	6/27 - 7 H 30 M	6/28 - 2 H 15 M	6/28 - 7 H 30 M
R4	6/28 - 7 H 30 M	6/30 - 0 H 00 M	6/29 - 7 H 30 M	7/01 - 0 H 00 M

	PRODUCT B1		PRODUCT B2	
RESOURCE	Scheduled Start	Scheduled Stop	Scheduled Start	Scheduled Stop
R1	6/25 - 4 H 40 M	6/26 - 2 H 10 M	6/27 - 4 H 40 M	6/28 - 2 H 10 M
R2	6/27 - 2 H 10 M	6/27 - 5 H 40 M	6/29 - 2 H 10 M	6/29 - 5 H 40 M
R5	6/28 - 5 H 40 M	6/30 - 0 H 00 M	6/30 - 5 H 40 M	7/02 - 0 H 00 M

TABLE 6.5   DETAILED DISPATCH LIST FOR RESOURCE R2 FOR THE TWO BATCHES OF PRODUCT A (A1 AND A2) AND TWO BATCHES OF PRODUCT B (B1 AND B2)

PRODUCT	OPERATION NUMBER	SCHEDULED START	SCHEDULED STOP
A1	020	6/24 - 0 H 40 M	6/24 - 4 H 15 M
A2	020	6/25 - 0 H 40 M	6/25 - 4 H 15 M
B1	020	6/27 - 2 H 10 M	6/27 - 5 H 40 M
A1	040	6/27 - 2 H 15 M	6/27 - 7 H 30 M
A2	040	6/28 - 2 H 15 M	6/28 - 7 H 30 M
B2	020	6/29 - 2 H 10 M	6/29 - 5 H 40 M

A closer examination of the failure of the MRP approach with respect to these two aspects is warranted.

Communicating Information   The typical dispatch list is ineffective as a communication system for two basic reasons. On one hand, the dispatch lists usually contain some details that the user already knows. Clearly, this type of information is of no value. On the other hand, much of the dispatch list information is either erroneous or is of no practical use. For example, consider the very precise start and finish times for various jobs that are normally part of the dispatch list information. These times are often stated down to the hour or minute. In reality, however, the job can be started only when the material becomes available. In addition, start and finish times are based on system data, scheduling rules, and processing and setup times that are often far from accurate. The shop floor supervisors and manufacturing personnel should be experts on which machines to use and the time required to process the products. They may even have special insights as to how to run the jobs that are not reflected by the dispatch list. As a result, experienced managers and operators may pay little attention to dispatch list information, especially in plants where expediting is common.

Only one piece of useful information is contained in the actual schedule generated by the planning system that the user really needs. This is the sequence in which jobs are to be run. It can be argued that in the relatively simple process just discussed, the only information that needs to be communicated to the work stations is the processing sequence. Evidently, something far simpler than a dispatch list is necessary for effective communication.

Controlling the Schedule   The techniques used in MRP II systems attempt to obtain information from all work stations and use this to control the process. But this approach fails for two reasons.

First, the MRP logic is based on the erroneous assumption that if every work station is instructed to adhere to the schedule, then the entire plant can stay on schedule. This argument again has common sense appeal. But disruptions to the schedule are inevitable and cumulative, and the argument falls apart in the interactive system of manufacturing operations. In an effort to reduce the effect of disruptions, management must establish buffers at each work station. This has already been demonstrated to be undesirable since it increases the lead time without providing protection for the entire system. Since individual buffers are often unable to compensate for significant deviations, management must resort to expensive alternatives such as the use of overtime in an effort to stay on schedule. (Remember that in synchronous manufacturing systems, most work stations do not have buffers. Thus, most work stations will experience the effect of every upstream disruption in the flow, making it virtually impossible for them to stay on schedule. But the *plant* will remain on schedule.)

Second, the task of accumulating and compiling all of the information necessary to establish control in an MRP system is a mammoth undertaking. Lack of education and discipline has typically taken the blame for the failure to perform this task effectively. But the almost universal lack of success in establishing tight control over the system indicates the existence of a more fundamental and serious problem. [33] In a system as interlocked and complex as a manufacturing plant, tight control through the micromanagement of all details is not feasible. Such attempts only result in a loss of credibility and control.

Simplifying the Control Problem   The basic problem of control is one of making sure that all work centers perform the right tasks in the right sequence. The simplest and most effective way to ensure that a work center does the right job is to make only that job available to the work center. After all, the job must be physically available to be processed. And if only the right job is available, then the work center is denied the opportunity to process the wrong job. Using this approach, the emphasis of control shifts away from the traditional method of trying to prioritize accurately the numerous available jobs. The new approach emphasizes limiting the quantity of material available on the shop floor to only what is needed.

The availability of material is the key to controlling the execution of the planned product flow. And material availability is largely a function of the quantity of material pumped into the system at the material release points. Thus, the material release points (the gateway operations) in the plant must be strictly controlled! To implement the DBR system, the material release operation needs to be provided with a detailed schedule—what products or materials are to be processed, in what quantity, and at what time. If this task is properly managed, the remaining problems in controlling the flow will be less troublesome. In most cases, the remaining work centers will simply process the material that is made available to them, as it becomes available.

In the DBR system, every batch that becomes available at a work center (except those with a time buffer) will either be on schedule or behind schedule. In either case, the material should be processed as soon as possible. Since lead times and work-in-process inventories have been greatly reduced in this system, work centers will rarely have more than one batch available to process. On those rare occasions, the dispatch list simply acts as a priority list.

In a simple linear process, control of the process is essentially reduced to control over the release of materials. Once the materials are released into the system, workers at the various resources simply need a priority list that they can consult when multiple orders are available. The recommended approach to controlling a simple linear process is somewhat similar to that used for assembly lines, where production is precisely controlled by the release of material to the line. The difference is that in assembly lines, individual stations do not receive any detailed instructions. They simply process the work as it is delivered.

In most plants, the overactivation of nonconstrained resources is the primary cause of material flow problems. But in simple linear processes, the use of time buffers and tight material control makes activation without utilization virtually impossible at most resources. As a result, in simple straight-line processes with DBR systems, it is not necessary to monitor the nonconstrained resources in great detail.

But most manufacturing environments are more complex than just a simple linear process, and control becomes a more difficult task. In most manufacturing plants, simply scheduling and managing the release of material is necessary, but not sufficient, to establish the desired degree of control over the production process. To maintain an effective DBR system, it may be necessary to develop additional control points within the plant.

Schedule Release Points In the DBR system, any point in the flow where a detailed schedule is necessary to maintain control over the product flow is called a *schedule release point*. It should be clear from the previous discussion that a detailed schedule is necessary whenever the availability of material is not sufficient to authorize the next work center to process the material. Schedule release points can be characterized as points where activation may occur without utilization. In most processes, control at a relatively few schedule release points eliminates the necessity of closely controlling all other points in the flow because there is no opportunity to overactivate.

Schedule release points, even in complex flows, are defined to exist at only four categories of points. These are (1) material release points, (2) CCRs, (3) divergence points, and (4) assembly points.

No matter how simple or complex the product flow, the material release point must be a schedule release point. Every point where material enters the system must be scheduled in order to control the material inputs into the system.

As previously discussed, effective control at CCRs is critical. Therefore, the CCRs of a process must also be schedule release points.

Detailed schedules are also required at divergence points. Divergence points, as illustrated in Figure 6.13, are points in the product flow where material can be processed into different products. Knowing how much of each product should be produced is critical to avoiding overactivation and the accompanying misallocation of material. While the timing of the jobs is still controlled by the availability of material, workers at each divergence point must be provided with a detailed list of what and how much of each product to produce, as well as the priority sequence for the products.

FIGURE 6.13   A DIVERGENCE POINT IN THE PRODUCT FLOW

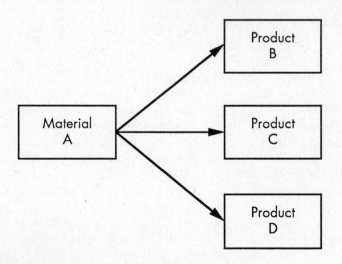

Assembly operations also require detailed schedules. At an assembly operation, as illustrated in Figure 6.14, the availability of a single part is not sufficient to begin the assembly process. By its very nature, many different parts are usually needed. The assembly workers will have to make sure that all of each required part is available. Thus, they need a detailed production

FIGURE 6.14   AN ASSEMBLY POINT IN THE PRODUCT FLOW

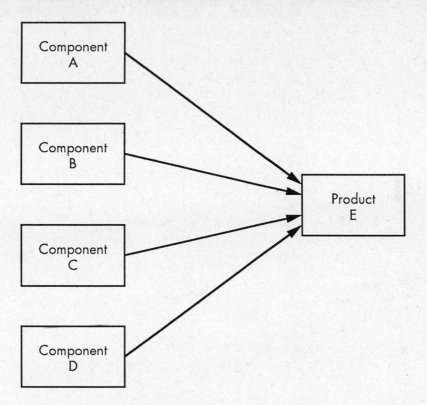

schedule so they will know what parts are needed at various points in time. Of course, the task of checking to make sure the required parts are available does not have to fall to the assemblers. Instead, this task may belong to the relevant stock room personnel. The specific assignment of responsibility is not important. However, it is critical to recognize the point in the flow where detailed control is necessary.

In a DBR system, a detailed schedule and strict management control of the process is needed only at the schedule release points—even for complex product flows. All other points (non-schedule release points) in the process require little control. The sequencing rules used at these work stations may be as simple as a "first-in, first-out" priority. The resulting simplification of communication requirements enhances the effectiveness of all communications and is a major advantage of the rope.

Scheduling Decisions and Tradeoffs   Once the schedule release points have been determined, the remaining issue in the rope is the actual development of the detailed schedule. The actual schedule is a function of the timing of material release, the process batch size, and the transfer batch size.

The timing for releasing material to any work center is determined by the length of time that must be allowed to process the material. The basic logic used for this purpose is the same as that used in standard MRP systems, except that the principles of synchronous manufacturing should be followed. The required manufacturing lead time is highly dependent on the choice of the process and transfer batch sizes.

The process batch size is determined by the MPS. And all relevant constraints and processing factors should have already been considered as the MPS was being derived. Since the rope is established in order to support the MPS, the process batches established by the MPS are considered to be set and not subject to change. Therefore, although the process batch sizes affect the detailed schedule, they are not variables at this point in the process.

The remaining major decision in the rope logic is the size of the transfer batch. As discussed in Chapter 5, the use of small transfer batches promotes smooth material flows, reduced levels of inventory, and shorter lead times. But using smaller transfer batches means that material must be moved between operations more often. When selecting the transfer batch sizes, the tradeoffs are between the benefits of shorter lead times and inventory reductions versus the costs of moving materials. When analyzing the tradeoffs, it must be remembered that the benefits of a rapid material flow are more than just inventory savings. The inevitable reductions in lead times bring forth new opportunities for increasing the firm's competitive advantage and market share.

A simple example can be used to illustrate the basic effects of transfer batch sizes. Figure 6.15 shows the production of a product as a simple linear process that has four processing steps, performed sequentially by resources R1, R2, R3, and R4. To simplify this example, no time buffers are indicated, although their inclusion would be a straightforward exercise. It is also assumed that the system is perfectly synchronized, so that when an order is released into the system, it moves through the system without waiting in queues at any resource. Setup times are negligible, and processing times per unit are

FIGURE 6.15   A PERFECTLY SYNCHRONIZED FOUR-STEP PROCESS

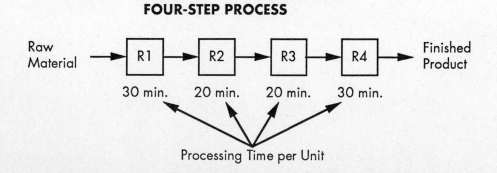

30, 20, 20, and 30 minutes per unit at resources R1, R2, R3, and R4, respectively. Suppose the MPS calls for a process batch of 30 units.

If the transfer batch is equal to the process batch, then the batch will require 15 hours of processing time at resource R1. The entire 30 units are then moved to resource R2 where another 10 hours of processing time is required, followed by 10 hours at R3, and 15 hours at R4. The batch is scheduled to complete processing 50 hours after the material is released into the system. This situation is illustrated in Table 6.6(a).

Now consider what happens if the transfer batch is reduced to 6 units, resulting in five batches of 6 units each. The first batch of 6 units is completed at R1 and moved to R2 to begin processing after only 3 hours. The last batch of 6 units is completed at R1 and moved to begin processing at R2 after 15 hours. The last batch completes processing at R2 after 17 hours. Following this last batch of 6 units through the process, we find that it completes processing at R3 and R4 after 19 and 22 hours, respectively. Thus, the entire order of 30 units is ready for shipping after only 22 hours from the release of materials into the system. The first batch of 6 units should also be tracked through the process, and found to be completed and ready for shipping in 10 hours. This situation is illustrated in Table 6.6(b).

TABLE 6.6   THE EFFECT OF TRANSFER BATCH SIZE ON MANUFACTURING LEAD TIME FOR THE PROCESS DESCRIBED IN FIGURE 6.15

(a) Process Batch = Transfer Batch = 30 Units

RESOURCE	START TIME	FINISH TIME
R1	00 Hrs	15 Hrs
R2	15 Hrs	25 Hrs
R3	25 Hrs	35 Hrs
R4	35 Hrs	50 Hrs

(b) Process Batch = 30 Units; Transfer Batch = 6 Units

RESOURCE	FIRST BATCH OF 6 UNITS Start Time	Finish Time	SECOND BATCH OF 6 UNITS Start Time	Finish Time	LAST BATCH OF 6 UNITS Start Time	Finish Time
R1	00 Hrs	03 Hrs	03 Hrs	06 Hrs	12 Hrs	15 Hrs
R2	03 Hrs	05 Hrs	06 Hrs	08 Hrs	15 Hrs	17 Hrs
R3	05 Hrs	07 Hrs	08 Hrs	10 Hrs	17 Hrs	19 Hrs
R4	07 Hrs	10 Hrs	10 Hrs	13 Hrs	19 Hrs	22 Hrs

In this example, reducing the transfer batch from 30 to 6 units results in the manufacturing lead time being reduced from 50 hours to 22 hours. This might result in delivery times that are cut in half. Furthermore, if a customer wanted quick delivery of a partial shipment, the first 6 units could be delivered in as little as 10 hours. Another important consideration is that using smaller transfer batches also yields a reduced level of WIP inventory in the system during the lead time period. The above analysis clearly shows the power of small transfer batches in a synchronized manufacturing environment.

A Drum-Buffer-Rope Illustration

Consider a plant that has a work cell that produces only two items, A and B. The two products are each processed through five resources, R1, R2, R3, R4, and R5. Each resource is available for 480 minutes per day. The projected customer demand, routings, setup times, and processing times for products A and B are all shown in Table 6.7.

Given the level of demand and the setup and processing times, it is evident that the CCR in this problem is resource R4. Therefore, two time buffers are needed, one in front of resource R4 and one before shipping. Suppose that the current manufacturing lead time for this process for either product A or product B is 3 weeks. A reasonable starting point for length of the time buffers is 3 days for each buffer. The resulting time buffer setup is graphically illustrated in Figure 6.16.

The next step is to establish the drum beat, i.e., the master production schedule for the process. The recommended procedure is to first convert the demand into a schedule at resource R4 (the CCR), and then use the CCR schedule to determine the MPS.

The process to derive the CCR schedule is illustrated in Tables 6.8 and 6.9. In Table 6.8, the daily schedule calls for R4 to process exactly one day's worth of demand for both products A and B. But close examination of the information in Table 6.8 indicates that resource R4 is assigned 510 minutes of work during each 480 minute work day. Thus, the trial schedule is not feasible. The schedule must be modified to bring it into line with the available capacity at R4. One simple way to do this is to double the process batch size for both A and B. The resulting schedule at resource R4 is described in Table 6.9.

The derived master production schedule (for the completed products) is shown in Table 6.10. This schedule reflects both the 3-day time buffer before shipping and the expected processing and setup times at resources R4 and R5. The differences between the original statement of demand and the derived MPS are conceptually summarized in Figure 6.17. Figure 6.17(a) demonstrates that the mere existence of demand does not presume any processing order

TABLE 6.7   CUSTOMER DEMAND AND ROUTINGS FOR TWO PRODUCTS, A AND B

(a) Customer Demand

DATE	PRODUCT	QUANTITY
6/26	A	10
6/26	B	5
6/27	A	10
6/27	B	5
6/28	A	10
6/28	B	5
.	.	.
.	.	.
.	.	.

(b) Routing File for Product A

OPERATION	RESOURCE	PROCESSING TIME PER UNIT (MINUTES)	SETUP TIME (MINUTES)
010	R1	25	15
020	R2	20	60
030	R3	20	00
040	R4	30	30
050	R5	20	00

(c) Routing File for Product B

OPERATION	RESOURCE	PROCESSING TIME PER UNIT (MINUTES)	SETUP TIME (MINUTES)
010	R1	25	15
020	R2	30	60
030	R3	25	00
040	R4	30	30
050	R5	40	00

for products A and B. The only requirement is that the quantities demanded are actually available as scheduled at the end of each day. But the derived master production schedule shown in Figure 6.17(b) reflects the realities of the manufacturing environment, specifically the capacity limitations of resource R4.

FIGURE 6.16   THE STRATEGIC PLACEMENT OF TIME BUFFERS TO PROTECT THE PROCESS DESCRIBED IN TABLE 6.7

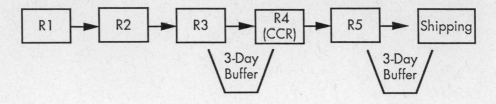

TABLE 6.8   ESTABLISHING THE DRUMBEAT FOR THE PROCESS DESCRIBED IN TABLE 6.7: CONVERTING CUSTOMER DEMAND INTO A SCHEDULE AT THE CCR (RESOURCE R4)

DATE	PRODUCT	OPERATION NUMBER	QUANTITY	RUN TIME (MINUTES)	SETUP TIME (MINUTES)
6/22	A	040	10	300	30
6/22	B	040	5	150	30
6/23	A	040	10	300	30
6/23	B	040	5	150	30
6/24	A	040	10	300	30
6/24	B	040	5	150	30

The schedule release points in this example are at material release (or resource R1) and resource R4 (the CCR). The critical issues for the material release function are what quantities of material to release and when to release them. The batch sizes have already been determined as 20 units of A and 10 units of B. Referring back to Table 6.7, it can be calculated that the total processing and setup time required for a batch of 20 units of product A is exactly 5 days. Similarly, the total processing and setup time required for a batch of 10 units of product B is 3 days 2 hours and 45 minutes. Using a total time buffer of 6 days, materials required for product A should be

**TABLE 6.9   ESTABLISHING THE DRUMBEAT FOR THE PROCESS
DESCRIBED IN TABLE 6.7: DERIVING A FEASIBLE
SCHEDULE AT THE CCR (RESOURCE R4)**

DATE	PRODUCT	OPERATION NUMBER	QUANTITY	RUN TIME (MINUTES)	SETUP TIME (MINUTES)
6/22-23	A	040	20	600	30
6/23	B	040	10	300	30
6/24-25	A	040	20	600	30
6/25	B	040	10	300	30
.	.	.	.	.	.
.	.	.	.	.	.
.	.	.	.	.	.

released (to resource R1) 11 days before an order is due, while the materials for product B should be released 9 days 2 hours and 45 minutes before an order is due. Since resource R4 is a CCR and a schedule release point, workers at this resource must know the processing priority for each work order arriving at its buffer. Since each work order is subject to disruptions that may delay its timely arrival to the time buffer, a simple "first-in, first-out" priority is not appropriate at R4. Since resources R2 and R3 are not schedule release points, a detailed schedule or sequence is not required. In fact, workers may simply be instructed to process work orders in the sequence that they arrive at the work center.

If it is assumed that all orders are due by the end of the day for each stated due date, then the material release schedule will be as shown in Table 6.11. An additional explanatory comment is appropriate here. The first two batches of material released in this schedule are both for product A. This is caused by the relative size of the lead times for the two products and the fact that there are no other orders already in the system. In the steady state demand mode indicated by Table 6.7, the materials for batches of A and B would be released alternately.

TABLE 6.10 **ESTABLISHING THE DRUMBEAT FOR THE PROCESS DESCRIBED IN TABLE 6.7: DERIVING THE MPS BASED ON THE FEASIBLE SCHEDULE AT THE CCR (RESOURCE R4)**

DATE	PRODUCT	QUANTITY
6/26	A	20
6/26	B	10
6/28	A	20
6/28	B	10
6/30	A	20
6/30	B	10
.	.	.
.	.	.
.	.	.

FIGURE 6.17 **A COMPARISON OF THE ORIGINAL SCHEDULE OF CUSTOMER DEMAND AND THE MPS**

(a) Original Schedule as Dictated by Customer Demand

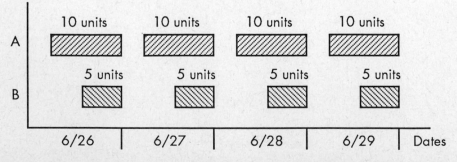

**FIGURE 6.17   A COMPARISON OF THE ORIGINAL SCHEDULE
OF CUSTOMER DEMAND AND THE MPS
(continued)**

(b) The Derived Master Production Schedule

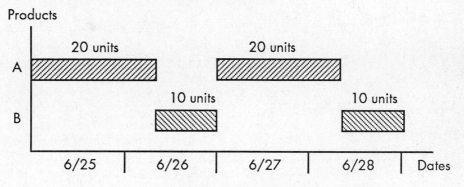

**TABLE 6.11   ESTABLISHING THE ROPE FOR THE PROCESS
DESCRIBED IN TABLE 6.7:
DERIVING THE MATERIAL RELEASE SCHEDULE
THAT SUPPORTS THE MPS**

PRODUCT	QUANTITY	RELEASE DATE
A	20	6/15 - 0 H 00 M
A	20	6/17 - 0 H 00 M
B	10	6/17 - 5 H 15 M
A	20	6/19 - 0 H 00 M
B	10	6/19 - 5 H 15 M
.	.	.
.	.	.
.	.	.

SUMMARY

In order to establish and maintain synchronized material flows in a plant it is necessary to properly plan and tightly control execution. The drum-buffer-rope system described in this chapter provides a practical and highly effective method for achieving synchronous flows in complex and dynamic manufacturing environments.

The drum provides a systematic approach to the problem of developing a master production schedule that is consistent with the constraints of the system. A thorough analysis of a plant's capabilities and the manufacturing environment within which it operates is used to identify the system constraints. The recommended procedure includes modifications to the derived MPS, which improves the quality of the resulting detailed production plan.

The time buffers are established to protect overall plant performance from the devastating effect of disruptions. Time buffers are required in only a very small number of strategic locations throughout a plant. The proficient use of time buffers can provide an optimum level of protection for a given level of WIP inventory.

The rope provides a simple and straightforward approach to controlling the production process. It reduces the problem of controlling the operation at each and every work center to one of establishing schedule release points only at strategic locations. At these points, schedules and discipline are required to maintain a smooth and timely product flow. The rope mechanism also reduces the problem of communicating the MPS requirements to the non-schedule release points since these work centers may be governed by simple priority sequencing rules such as "first-in, first-out."

QUESTIONS

1. What is the drum beat?
2. What are the factors that influence the setting of the drum beat?
3. Compare the ropes used in the kanban system with the concept of rope as developed in this chapter.
4. Compare and contrast time buffers and stock buffers. Identify situations where each is desired or needed.
5. What are some of the factors that influence the size and location of (a) time buffers and (b) stock buffers?
6. Using a schedule performance curve analysis, what conclusions might you reach based on the following data sets:

a. Order #	Customer Due Date	Scheduled Completion Date
101	4/10	4/11
105	4/15	4/11
107	4/18	4/11
113	4/20	4/21
118	4/24	4/21
124	4/29	4/21

b. Order #	Customer Due Date	Scheduled Completion Date
101	4/10	4/8
105	4/15	4/12
107	4/18	4/15
113	4/20	4/16
118	4/24	4/19
124	4/29	4/26

PROBLEMS

1. Using the following information, identify the schedule release points and the location of the time buffers:

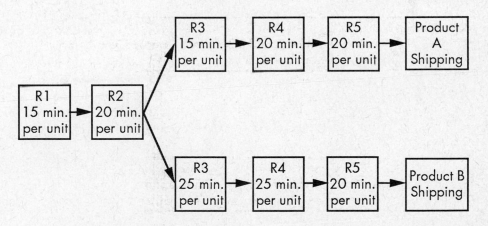

Setup Time at R3, R4, and R5 = 60 Minutes

Each Resource Has Available Capacity of 480 Minutes per Day

Daily Demand: Product A = 10 Units   Product B = 10 Units

2. Refer back to the information in the previous question. Make whatever assumptions are necessary and establish a workable drum-buffer-rope system for the process.

7

Focusing the Improvement Process

MARKET-CONSTRAINED VERSUS PRODUCTION-CONSTRAINED FIRMS

Constraints limit the performance and profitability of all organizations. To fully tap a firm's potential and improve profitability, it is first necessary to recognize the type of constraints that actually limit the organization's performance. Manufacturing firms can generally be classified as either market-constrained or production-constrained. Market-constrained firms have the ability to produce more product than the market is willing to buy under current competitive conditions. Conversely, production-constrained firms are unable to consistently satisfy the available market demand for products. The appropriate approach to improving performance depends upon whether the firm is market-constrained or production-constrained.

Market-Constrained Firms

Market-constrained plants have excess production capacity that cannot be converted into throughput under current market and operating conditions. As a result, strategies designed to increase output are worthless. Moreover, strategies designed to increase productivity by conventional cost-cutting procedures that focus on reductions in direct labor are also of limited value. Such actions may even result in lost capacity, which may transform resources into bottlenecks or CCRs, and in turn may disrupt the production process and adversely affect throughput. The implementation of labor-based cost-cutting programs is not the recommended strategy for market-constrained firms. Such strategies, if successful at all, normally result in marginally small improvements at the expense of restricting future plant performance and growth.

The preferred strategy for market-constrained plants is to increase the demand for its products. This is accomplished by enhancing the firm's competitive edge factors—producing high-quality products, offering superior delivery performance, and being a low-cost producer. (Cost reductions are best accomplished through actions that decrease inventory and improve quality.) Assuming a competent sales staff, the inevitable result of enhancing these three aspects of competition is that market demand will increase. And since the firm has excess capacity, some additional throughput can be produced for virtually the cost of the required raw materials.

As discussed in Chapter 2, the key to improving the three competitive edge variables lies in developing a more highly synchronized product flow. In addition to potentially higher levels of throughput, additional benefits of a synchronized product flow also include lower levels of inventory and reduced operating expenses. As a firm develops a synchronized product flow and increases its competitive advantage in the marketplace, it also becomes more profitable. But as the demand for the firm's products increases, the plant's excess capacity dwindles. Eventually, the market-constrained plant is likely to be transformed to a production-constrained plant.

Production-Constrained Firms

Production-constrained plants have trouble producing sufficient throughput to keep pace with market demand. Thus, the firm's primary concern is no longer marketplace competitiveness. Issues such as improving product quality and cutting costs become less important. The emphasis naturally shifts to increasing the level of output of the plant to match the already existing demand. That is, attention is now focused on increasing the flow of product through the plant.

Production-constrained plants can fall short of market expectations for throughput in three different ways. Such plants (1) may have trouble producing enough total product to meet total demand, (2) typically have long lead times because of production backlogs, and (3) often cannot consistently meet promised shipping dates.

In the short run, perhaps the most significant of these three problems is the first one listed—the average quantity of product that can be produced is less than the average amount demanded. If a plant cannot produce enough product to satisfy the available customer demand, then it is losing potential sales and profit. When this situation occurs, it is usually because of the existence of bottlenecks or poorly managed CCRs in the manufacturing process.

In the long run, lengthy lead times and missed shipping dates can be just as harmful to company profitability as limited production levels. This is because long lead times and missed or late shipments can destroy a firm's competitive advantage, especially in a highly competitive marketplace.

The solution for missed shipping dates lies in the establishment of a drum-buffer-rope system coupled with a realistic order-promise system. This type of synchronized system provides excellent protection for plant throughput.

The solution for the long lead time problem also lies partially in the establishment of a synchronized product flow. However, this only ensures that the manufacturing lead time is relatively short. The complete solution must include a reduction of any existing backlog of customer orders so that backlogged orders are released to the shop floor without a lengthy waiting period. Such a reduction of backlog in a production-constrained firm will require at least a temporary increase in throughput for the plant.

The complete solution for improving performance for production-constrained firms includes both the implementation of a synchronized product flow and an increase in throughput for the plant. The drum-buffer-rope logistical system has already been presented as a way to synchronize the product flow. In the next section, our attention shifts to methods for increasing throughput for the plant.

TECHNIQUES FOR INCREASING THROUGHPUT

When market demand is not a constraint on the amount of throughput that can be generated by a plant, the limiting constraint or constraints must be logistical, managerial, behavioral, material, and/or capacity. Despite the likely existence of all of these constraints in a manufacturing plant, in most cases the primary constraint on throughput in production-constrained plants is capacity. Thus, our discussion should focus on those few resources that effectively control the throughput of the entire plant, the capacity-constrained resources (CCRs).

CCRs, by definition, have a limited amount of available capacity relative to the load required to support the planned product flow. This is especially true if a CCR is also a bottleneck. Therefore, managers must ensure that the CCRs are utilized to their fullest capability in support of the product flow. There are several courses of action that can be pursued in order to better utilize the CCRs. [1, 2]

Eliminate Periods of Idle Time

Idle time occurs at all resources, even the most critical bottlenecks in the plant. Idle time at a resource may be classified as either expected or unexpected. Expected idle time occurs when a resource is being unproductive by design. If desired, this type of idle time is usually quite easy to remedy. Unexpected idle time results when a resource is supposed to be engaged in productive work, but circumstances prevent it. This category of idle time can be considerably more difficult to remove from the system.

Expected idle time may occur repeatedly in a plant at regular time intervals because no workers are available to run the equipment. In many plants, workers routinely leave critical pieces of equipment unattended and idle during parts of the workday. For example, it may be a common practice to completely shut down a process during lunch, coffee breaks, and between shifts. Of course, such practices result in a loss of valuable capacity. Management must make sure that workers' schedules are designed so that CCRs are continuously operational. Instituting such changes alone may result in as much as 20 percent additional capacity at a critical resource.

Unexpected idle time occurs at irregular time intervals and may be caused by any number of random events. Absenteeism, a lack of materials or tools, or malfunctioning equipment are the major reasons for unexpected idle time.

In some operations, the absence of a key employee may result in a critical resource being unmanned for one or more days. In order to prevent the loss of valuable capacity, management must have contingency plans for this situation. Cross-trained workers who can step in and keep the CCR operating and a trigger mechanism to initiate the action are necessary.

Poor planning or mismanagement of the product flow or work schedule can result in a CCR with no material to process. The CCR is ready to work but is starved for work. Conversely, the work may be available, but because of a lack of tooling or necessary supplies, the resource is unable to commence processing. It is management's job to minimize these types of disruptions at critical resources.

Downtime due to malfunctioning equipment is a problem common to all manufacturing plants. This naturally results in lost capacity. The best approach to reducing the frequency of unexpected equipment failures is through a consistent program of preventive maintenance.

Reduce Setup and Processing Time Per Unit

Available time at a resource can be categorized as either setup time, processing time, waste time, or idle time. It is possible to increase the effective capacity of a resource by reducing the required processing time per unit, reducing waste time, or cutting the total setup time.

Historically, great emphasis has been placed on improving the processing speed of the equipment and workers in manufacturing plants. The numerous available procedures do not need to be discussed here. However, it is appropriate to emphasize that these efforts to improve processing time capabilities should focus on the bottlenecks/CCRs in the system. Significant gains in systemwide performance can only be recognized if processing improvements are made at these critical resources.

Waste time is not always easy to identify. However, any reduction of waste time will lead to a proportional increase in available capacity, which may be used to help generate throughput.

If a resource has no excess capacity, then any reduction in the total required setup time can be converted into additional available processing time. Total setup time at a resource can be reduced in two basic ways.

One approach to decreasing total setup time is to reduce the total number of setups that are performed in a given period of time. (This approach is inappropriate for non-CCRs since they already have excess capacity and larger batch sizes would disrupt the smooth flow of materials.) Reductions in the number of setups may be achieved through the development of optimum (not EOQ) batch-sizing policies at the critical resources. It may be possible to reduce the total number of batches simply by combining batches and avoiding the splitting of batches often caused by expediting "hot" orders. If so, then total setup time can be reduced and processing capacity increased. However, remember that there are limitations to how large a batch should be. The throughput requirements and the necessity of maintaining a smooth and continuous flow of products according to a synchronized plan must be considered.

A second approach is to reduce the time required for each individual setup. This may be accomplished through either batch sequencing or setup engineering.

In numerous processes, the setup time for a particular batch of product is dependent upon the processing requirements of the previous batch of products. For example, in heat treat processes, the temperature requirements for different products may vary greatly. The setup time consists primarily of the time required to let the furnace cool down or heat up to the specified temperature. Thus, in scheduling batches of products for a heat treat operation, the sequence of the batches plays a huge role in the setup time requirements for the furnace. Ideally, to maximize output at a heat treat operation, batches of products that have approximately the same temperature requirements should be sequenced together whenever possible.

The setup time per batch can also be reduced through setup engineering. The steps required to complete a setup should be analyzed in order to find ways to reduce the total setup time. (One good approach is to videotape the setup operation and have the operator suggest improvements.) The Japanese have become most proficient at reducing setup times through such process engineering. One basic methodology is to divide the setup tasks into activities that are internal and external to the process. Internal setup activities can only be performed while the process is not operating and producing products. External setup activities are those that can be performed while the resource is running. In fact, external setup activities for the next batch often can be completed while the current batch is being processed. Activities such as securing the necessary tooling and materials are examples of external setup activities. Another approach is to redesign the setup procedures, equipment, dies, or tooling in such a way as to make the setup or changeover activity easier to perform. There are numerous specific guidelines for reducing setup times. But a discussion of those guidelines is beyond the intended coverage of this text.

Improve Quality Control

Quality control is extremely critical on parts that are processed by a CCR. It is important that CCRs not work on defective parts. It is also important that parts already processed by a CCR do not become defective, requiring the product to be either scrapped or reworked. It is management's task to devise and implement policies and procedures that support these guidelines.

Logically, CCRs should not work on any parts that are already defective. When a CCR works on material that is defective and not fit for throughput, then the time expended at the CCR to process the defective unit has been wasted. Since the available capacity at the CCR is already a valuable and scarce commodity, this waste cannot be tolerated. Therefore, the quality of materials utilized at the CCR must be within process and product specifications. If necessary to guarantee the required level of quality, inspection should be placed immediately before the CCR, i.e., at the material inbound stock point for the resource.

Assuming that acceptable parts are being produced by a CCR, the next step is to develop procedures which try to minimize the likelihood that these parts will become defective and scrapped because of a later operation. The rationale is that any part processed by a CCR that must be scrapped results in lost throughput. This also means wasted capacity for that CCR. Again, it is management's responsibility to establish procedures designed to prevent the CCR-processed parts from becoming defective after they leave the CCR. This is just as important as ensuring that a CCR does not work on defective materials in the first place.

Suppose, despite all precautions, that a CCR-processed part becomes defective at a downstream operation. The question then becomes one of whether or not the part can or should be reworked. If it is possible to rework the part, the question becomes whether the part requires rework at the CCR. If rework is required, another question is how much CCR rework time is required. If the amount of time required to rework the part at the CCR exceeds the normal processing time per unit, then it may be more productive to simply scrap the part. This is especially true if the CCR in question is a bottleneck. If the amount of rework time at the CCR is less than the normal processing time for the part, then another consideration is whether the part is likely to be scrapped after the rework is performed. The clear implication is that management first must understand the value of processing time at the CCR.

Then the appropriate quality inspection, scrap, and rework policies for parts can be logically derived.

Reduce the Workload

Management should carefully examine the parts that are processed by CCRs. A thorough analysis often reveals that some of the materials processed by CCRs should not even be on the work schedule. For example, in many plants it is possible to find materials being processed through CCRs for which there is no market demand. Meanwhile, critical materials wait in the queue. Since CCRs have limited processing capacity, valuable capacity is being misallocated. It is clear that in many cases, much valuable processing capacity at some CCRs can be made available if only materials that contribute to throughput are processed. A top priority of management should be to ensure that only products which can be converted into throughput are processed.

Another situation that results in needlessly excessive workloads on CCRs occurs because the engineering or customer specifications of most products demanded by the market change over time. Sometimes these changes are slow in getting implemented at the shop floor level. As a result, products may continue to be processed by specific resources, even though processing is no longer necessary. If a resource that is no longer a required step in the product routing is a CCR, then the CCR is needlessly overloaded.

Management should consider taking positive action to redesign processes or products in order to reduce the number of parts that must use a CCR. In many cases, more effective communication with customers or suppliers can provide information that will allow modifications to be made in the product or the process. If these potential modifications result in reduced demand on a CCR, this is the same as increasing available capacity at that resource.

Another alternative is to offload work to other machines that can perform the same function as the CCR. In some cases, older pieces of equipment are already available in the plant but are simply not used. The reason they are not utilized is often that they are slower and, therefore, not as cost effective. Despite what the local cost calculations indicate, if such machines can provide extra capacity at a CCR, they should be utilized.

Another way to offload capacity is by subcontracting the work to another manufacturer. Even though the external cost of performing the work probably exceeds the internal processing cost, subcontracting may add extremely valuable CCR capacity to the system. The payback for offloading critical resources can be increased throughput and improved on-time delivery performance for the entire system.

Purchase Additional Capacity

The firm may want to consider purchasing new or additional equipment, hiring additional workers, or scheduling overtime. These alternatives are commonly used techniques to increase capacity. However, these alternatives should be considered as measures of last resort.

The purchase of additional equipment to increase capacity and throughput is usually a very expensive and tricky exercise. First of all, one must be sure that the new equipment is not targeted for use at a non-CCR. By definition, non-CCRs already have excess capacity, and any money spent to increase capacity at such resources is basically wasted. If the equipment is for use at a CCR, that equipment is likely to be quite costly. Especially in capital-intensive plants, CCRs are often characterized by processes that require expensive machinery. The high cost of the equipment is often a major reason why some resources are CCRs in the first place.

Hiring additional workers is also a costly proposition. However, if the workers are hired to run a second or third shift, this solution is probably less costly than purchasing additional equipment, which may also require additional workers. This alternative also usually provides the firm with greater flexibility since workers can be cross trained or used to perform multiple tasks. Using overtime is a less permanent and less risky way to increase capacity at a resource than hiring additional workers or purchasing new equipment. However, the use of overtime does carry with it the high cost of premium wages.

Management must understand that money spent to increase the capacity of resources should be targeted for CCRs. And buying additional equipment or increasing the scheduled number of man-hours for these resources should be seriously considered only after all of the previously discussed techniques for increasing throughput have been exhausted. Unfortunately, the purchase of additional capacity is often the first or only alternative considered by less than fully competent management. If so, the firm usually ends up paying handsomely for capacity that could have been made available for little or no cost.

UTILIZING THE BUFFER TO IMPROVE PERFORMANCE

Most managers fully understand how important it is to continually improve the manufacturing processes in their plants. But in a typical manufacturing plant, the number of improvements that can be attempted is almost unlimited, and it is not feasible to start making changes everywhere at once. The problem is determining which improvements will yield the greatest payback in terms of overall plant performance. Fortunately, help is available.

A major advantage of the drum-buffer-rope (DBR) logistical system of synchronous manufacturing is that it can help managers accurately focus their process improvement activities. The time buffers in a DBR system can be used to help identify the causes of disruptions that affect the planned product flow through the plant. [5, pp. 116–131] Once the various causes of disruptions are known, plans can be developed to improve the process by eliminating the causes of significant disruptions. The process improvement activities can be prioritized according to the significance of the disruptions that occur at the time buffers. In the remainder of this chapter, procedures are described that indicate how to manage the time buffers and how to use the buffers to focus the firm's process improvement efforts.

Analyzing the Time Buffer

As discussed in Chapter 6, the primary purpose of time buffers is to protect the throughput of the system from the inevitable disruptions that plague manufacturing plants. The ensuing analysis of how well a time buffer is working will lead directly to the development of a procedure to identify the most significant sources of problems in the process.

Visualizing the Time Buffer   To illustrate the analysis, a weekly work schedule for a CCR is presented in Table 7.1. This schedule includes information indicating a work order identification number, the part to be processed, the number of units to be processed, and the number of CCR hours necessary to process the order. The work sequence priority is indicated by the work order identification number. This schedule provides the basic information necessary to determine the *planned* content of the time buffer at the CCR on any given day. For example, if the CCR is to have a 4 day time buffer, then on Monday morning, all work that is scheduled for Monday, Tuesday, Wednesday, and Thursday is planned to be in the work queue in front of the CCR.

For ease of analysis, the planned content of the 4 day time buffer for Monday morning for the CCR identified in Table 7.1 is presented in a block format in Figure 7.1. The horizontal axis simply marks each scheduled workday for the CCR. The vertical axis indicates how the various work orders are scheduled during the available 8 hours of each work day. This format makes it easy to visualize the CCR schedule for the duration of the time buffer, and also facilitates the analysis of the actual versus the planned content of the time buffer.

Holes in the Time Buffer   The purpose of time buffers is to protect throughput and shipping schedules from the disruptions of the manufacturing environment. In the drum-buffer-rope logistical system, slack time is built into the process only at the time buffers. As a result, a single disruption at any one of most resources in the plant will throw the planned product flow off schedule. With no slack time built into the process except at the time buffers, there is little chance that the planned flow can be put back on schedule before the affected orders reach the buffer. Thus, the timely arrival of material at the time buffers can be adversely affected by disruptions at almost any point in the process.

TABLE 7.1   WEEKLY WORK SCHEDULE FOR A CCR

DAY	WORK ORDER ID NUMBER	PART ID	QUANTITY TO BE PROCESSED	CCR PROCESSING HOURS REQUIRED
Monday	101	C	100	3
	102	A	40	2
	104	E	30	1
	106	G	20	2
Tuesday	107	A	60	3
	108	C	100	3
	110	B	50	2
Wednesday	111	F	25	1
	112	G	30	3
	113	D	10	2
	115	E	60	2
Thursday	116	B	75	3
	117	A	20	1
	120	F	50	2
	121	G	10	1
	122	E	30	1
Friday	124	B	25	1
	125	H	90	3
	127	A	20	1
	128	D	15	3

Since disruptions are common occurrences in manufacturing plants, the actual content of the time buffers will typically be less than the planned content because of orders arriving late at the buffer. When an order is planned to be in the buffer but is missing, this is referred to as a hole in the buffer. Figure 7.2 illustrates this situation for the 4 day time buffer at the CCR previously described. In Figure 7.2, the buffer has six holes, with orders 110, 113, 115, 116, 120, and 121 currently not found in the buffer.

FIGURE 7.1   THE PLANNED 4 DAY TIME BUFFER
AS OF MONDAY MORNING FOR THE CCR
DESCRIBED IN TABLE 7.1

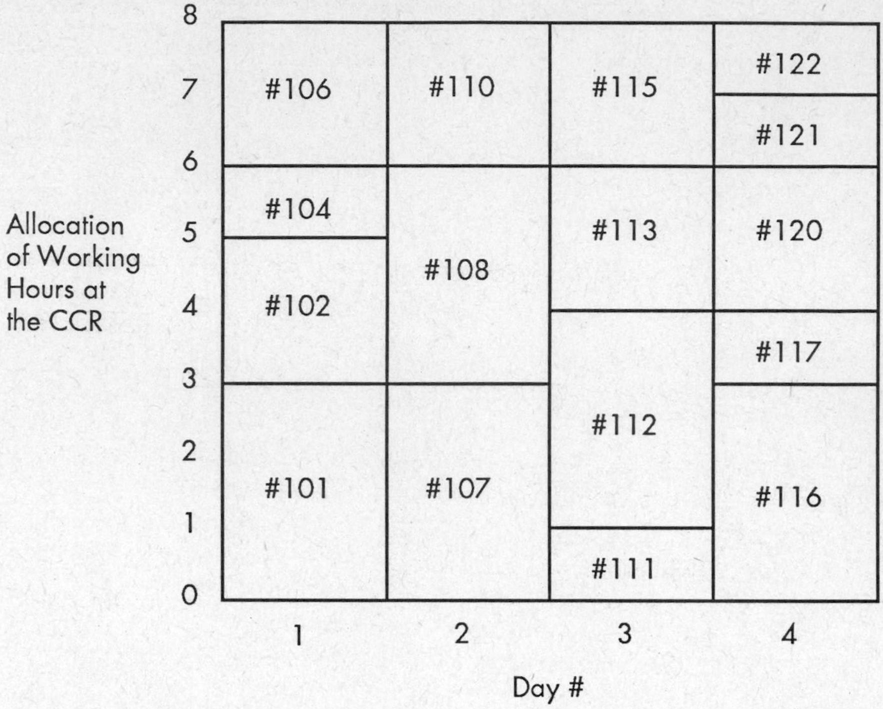

The situation in Figure 7.2 can be assessed very quickly. Since the CCR maintains a 4 day time buffer, 32 hours of work are planned to be available. But in actuality, since missing orders 110, 113, 115, 116, 120, and 121 account for a total of 12 scheduled hours of work, there are only 20 hours of work in the queue on Monday morning.

Of the six delinquent work orders, order 110 has the highest processing priority and is examined further. Processing for order 110 is scheduled to begin in 14 work hours and last for 2 hours. If order 110 does not arrive at the buffer within the next 14 work hours, then the CCR schedule will be affected. Since other work is available in the buffer, the CCR will not be forced to shut down. However, if order 110 is not processed on schedule at the CCR, this may also cause delays at other downstream operations or at shipping. Moreover, the schedule change may adversely affect CCR productivity if setup times at the CCR are dependent upon the processing sequence.

FIGURE 7.2   HOLES IN THE PLANNED 4 DAY TIME BUFFER AS OF MONDAY MORNING FOR THE CCR DESCRIBED IN TABLE 7.1

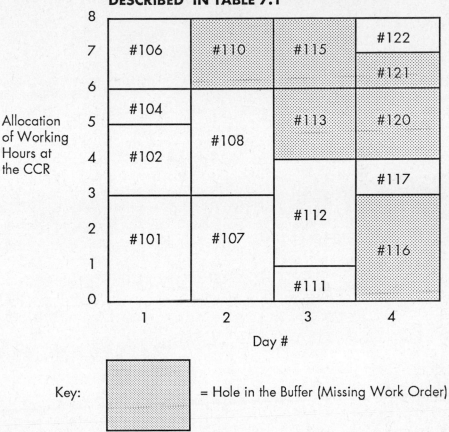

Key: = Hole in the Buffer (Missing Work Order)

Comparing Actual and Planned Time Buffer Content   In order to lay the groundwork for a meaningful analysis of the time buffer, it is necessary to develop the concept of *time buffer content profiles*. These content profiles indicate the approximate percentage of planned work that is, on average, actually present in a buffer for each time period covered by the buffer. For example, in the situation illustrated by Figure 7.2, all of the work planned for the first day (Monday) was actually in the buffer. Thus, the percentage of planned work in queue for Monday is 100 percent. For Tuesday's planned work, order 110, which represents 2 hours of work, is missing. This hole in the buffer means that only 75 percent of the work planned for Tuesday is available. Likewise, the percentages of planned work available for Wednesday and Thursday are 50 percent and 25 percent, respectively. Suppose for a moment that these calculated percentages are perfectly typical of the average percentages of planned work available for the 4 day time buffer. In that case, these percentages would constitute a content profile for the 4 day time buffer, which is illustrated in Figure 7.3.

A relevant question at this point is, "What should the time buffer content profile look like?" This question can be answered by considering several related issues.

First, consider what it means if the time buffer is always completely full. If a buffer is always full, the time buffer content profiles are all 100 percent. This means that there are no disruptions significant enough to make the actual product flow deviate from the planned product flow. In such a situation, a time buffer is not needed to protect the product flow and should be eliminated. Eliminating or decreasing unnecessarily large time buffers reduces lead time, inventory, and operating expense for the entire plant and enhances the firm's competitive edge. Clearly, it is neither likely nor desirable that all of the time buffer content profiles be 100 percent.

On the other hand, the time buffer content profile must be sufficient to achieve the buffer's primary purpose of protecting throughput. Thus, the

FIGURE 7.3   CONTENT PROFILE BASED ON THE PERCENTAGE OF WORK AVAILABLE ON MONDAY MORNING IN FIGURE 7.2

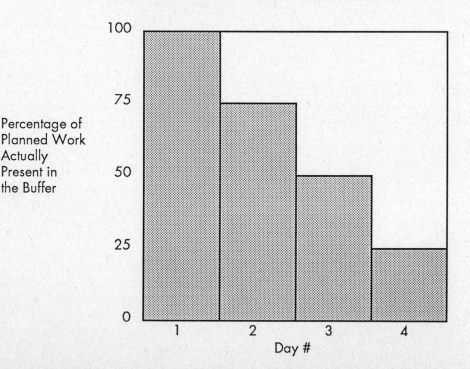

material scheduled for processing during the first portion of the time period covered by the buffer should be available almost all of the time. In our previous example, which has a 4 day buffer, this would mean that the time buffer content profile for the first day should be close to 100 percent. Admittedly, a content profile exactly equal to 100 percent for the first period of the buffer would be ideal from the scheduling perspective. But the cost of achieving 100 percent protection is likely to be prohibitively high, and such perfection probably cannot be guaranteed anyway. A content profile of about 95 percent to 99 percent for the first period of the time buffer is a reasonable and cost-effective goal in most cases.

The concept of acceptable content profiles for the remaining portions of the time buffer is quite variable. It is normal and acceptable for a majority of the material planned for the last period of the time buffer to be missing. Furthermore, the content profiles for the middle periods of the time buffer may fall almost anywhere between the extremes of the first and last periods, increasing as the scheduled processing time draws nearer.

The time buffer content profile depicted in Figure 7.3 would be considered excellent, especially since the first period has a content profile of 100 percent. But even if the content profile for the first period had been 95 percent instead of 100 percent, the overall time buffer content profile would normally be considered acceptable and sufficient enough to protect the throughput from all but the most severe disruptions.

The time buffer content profile can be a useful tool in properly managing the buffers and the planned product flow. To illustrate how this tool is used, we will consider three different unsatisfactory time buffer content profiles and discuss the steps that should be taken to correct the situations.

The content profile in Figure 7.4(a) indicates that the time buffer is too full. The throughput is overprotected, and the cost of this excessive protection is unnecessarily high lead time, inventory, and operating expense, which lead directly to a deterioration of the firm's competitive edge. The time buffer should be cut to the point where only the first period of the time buffer is completely, or nearly, full. In this case, the buffer can be cut from 4 days to 2 days, and the throughput is still well protected. Figure 7.4(b) shows the time buffer content profile after the buffer has been reduced from 4 to 2 days.

Figure 7.5(a) shows a time buffer that is inadequate to protect the system throughput. The first period of the time buffer has a content profile percentage that is too low. The content profile of the second and third days also appears to be somewhat low. The immediate solution in this situation is to increase the level of protection by increasing the duration of the time buffer. For example, the buffer can be increased from 4 to 5 days. Figure 7.5(b) indicates one possible scenario of extending the buffer to 5 days. It shows a content profile for the first day that is now close to 100 percent, an acceptable level of protection. The process here should be to extend the buffer by one period and then construct a new content profile. If the level of protection is still

FIGURE 7.4   TIME BUFFER CONTENT PROFILE AND RECOMMENDED ACTION: EXCESSIVE BUFFER

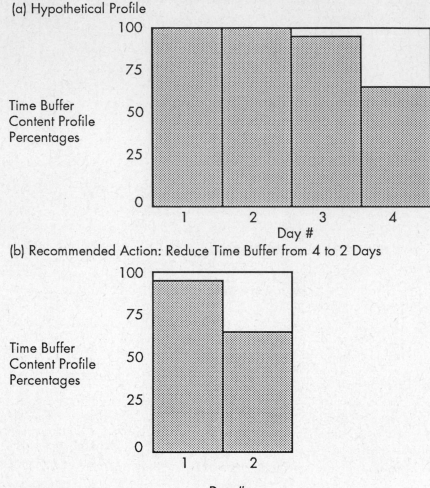

(a) Hypothetical Profile

(b) Recommended Action: Reduce Time Buffer from 4 to 2 Days

not adequate, then the procedure of extending the buffer should be repeated until a satisfactory solution is obtained.

Figure 7.6 illustrates a situation where the actual content of the time buffer includes unplanned material. This indicates that materials are arriving at the time buffer ahead of schedule. In the drum-buffer-rope system, this should not occur. This type of content profile suggests that material is being prematurely released into the system. The time buffer content profile has pinpointed a problem that probably can be traced back to the gateway

FIGURE 7.5   TIME BUFFER CONTENT PROFILE AND RECOMMENDED ACTION: INADEQUATE BUFFER

(a) Hypothetical Profile

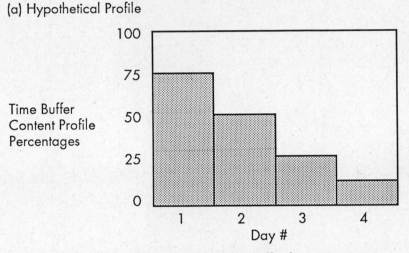

(b) Recommended Action: Increase Time Buffer from 4 to 5 Days

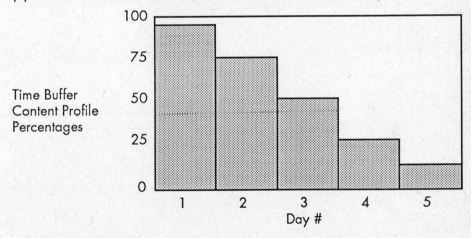

operations. The recommended solution includes additional education and discipline for those who control the material release function.

Implementing a Process of Continuous Improvement

The previous section illustrated how the time buffer can be used to identify when problems in the planned product flow exist. Now we shall see how

**FIGURE 7.6   TIME BUFFER CONTENT PROFILE
CONTAINING UNPLANNED MATERIAL**

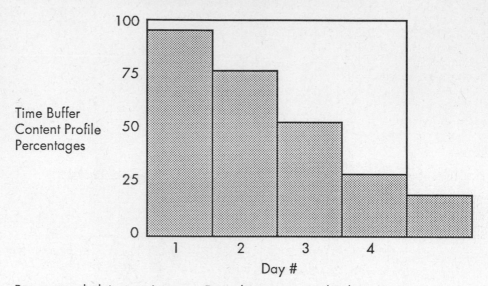

Recommended Action: Increase Discipline at Material Release Points

the time buffer can be used to help pinpoint the major causes of disruptions. Once the causes of disruptions are identified, improvement activities for the process can be prioritized so as to have the greatest positive impact on the plant.

The 80–20 Rule   In most systems, whether natural or manmade, an interesting phenomenon can usually be observed. Regardless of the characteristic of interest, there is usually a relatively small number of significant items and a much larger number of trivial ones. This phenomenon is known as the 80–20 rule. More specifically, this rule states that approximately 20 percent of the items in a system account for 80 percent of the characteristic of interest. Illustrations of this principle abound in everyday life. For example, in the United States, the ten largest states in the United States (20 percent of the total of 50) contain approximately 80 percent of the total geographic area. It is also approximately true that the ten most populous states have about 80 percent of the total population. A colleague with five children once joked that in any given year it seems as if 20 percent of his children account for 80 percent of his problems.

The 80–20 rule is very applicable to manufacturing environments. For example, experienced inventory managers are well aware that 20 percent of the SKU items in stock account for about 80 percent of the annual dollar inventory value in most systems. Likewise, 80 percent of the problems with raw materials in a plant typically can be attributed to 20 percent of the vendors. In a similar manner, the 80–20 rule can also be applied to sources of disruptions in a plant. That is, roughly 80 percent of the problems in a manufacturing process are usually attributable to about 20 percent of the work centers or vendors. Most of the really significant problems can be traced to a relatively few sources.

Improvement efforts in our plants should be guided by the 80–20 rule. It is of little value to make improvements in areas that yield only marginal benefits. The most desirable improvements are those where the derived benefits are relatively large. Thus, managers must identify those sources that account for the largest number and the most significant disruptions and strive to improve those areas first.

Deriving Individual Disruption Factors   The starting point for analyzing disruptions is the time buffer itself. The suggested procedure is simple and straightforward. First, the source (work center or vendor) of each disruption must be identified. (It is efficient to consider only significant disruptions.) Once this is accomplished, then an individual disruption factor is determined for each hole that appears in the time buffer. This disruption factor is based on the relative amount of damage done to the buffer.

When analyzing disruptions, one must first find out why the material is late in arriving at the time buffer. If the source of the problem is a vendor, the purchasing department can quickly identify the offender. In the case of disruptions at work centers, the source can often be identified by simply walking around and observing the process or talking to the supervisors or workers. Alternatively, the source of the problem can be tracked down by using information from the production control system used to monitor the progress of work orders. In some cases, creative detective work may be required to find the true source of the disruption. But this information is a necessary prerequisite to implementing a top-notch system of process improvement.

Every significant hole that appears in a time buffer is identified and attributed to a specific disruption source. The hole is closely monitored until the missing material finally arrives at the buffer, at which point the damage to the buffer can be assessed. The damage to the buffer caused by the disruption is a function of two variables. One variable is the size of the buffer hole, denoted by Z, which indicates the amount of processing time at the next resource represented by the late material. The other variable, denoted by L, measures how late the order is when it finally arrives at the buffer. Clearly, the greater the hole caused by the missing order and the later the order is, the greater is the damage to the time buffer. These two variables can be used to quantify an individual disruption factor for each hole in the buffer. The disruption factor may be quantified in many different ways. One such possible formula can be expressed as follows:

$$Individual\ disruption\ factor = Z \times L$$

Table 7.2 indicates the relative size of disruption factors calculated for four different holes that appear in a time buffer. The first disruption causes a half-day hole in the buffer and the material arrives at the buffer 1 day late, resulting in an individual disruption factor of .5. The second disruption causes a 1 day hole in the buffer, and the material also arrives 1 day late. Clearly, the second disruption is more severe than the first. In both cases, the missing material is 1 day late. But since the size of the hole caused by the second disruption is twice as large as the hole caused by the first disruption, the individual disruption factor for the second disruption is assigned a value two times larger (1.0) than the first disruption. The third disruption causes a half-day hole in the buffer, but the material arrives 2 days late. The derived individual disruption factor is 1.0, exactly the same as the disruption factor for the second disruption. The fourth disruption causes a 1 day hole in the buffer and is 2 days late, yielding a disruption factor of 2.0. Thus, the fourth disruption may be considered to be approximately four times more severe than the first disruption and twice as severe as the second and third disruptions.

TABLE 7.2   CALCULATION OF FOUR DISRUPTION FACTORS

DISRUPTION #	PROCESSING TIME (Z) (DAYS)	DEGREE OF LATENESS (L) (DAYS)	DISRUPTION FACTOR (Z X L)
1	.5	1	.5
2	1	1	1.0
3	.5	2	1.0
4	1	2	2.0

Prioritizing Improvement Activities   The recommended procedure for identifying the most significant sources of disruption in the plant is to derive a cumulative disruption factor for each source. A cumulative disruption factor is a single number, calculated for each work center or vendor, which quantifies the relative damage inflicted on time buffers over a given period of time because of disruptions originating at that source. The procedure used to determine the individual disruption factors is repeated for every significant hole that appears in every time buffer in the entire plant. These individual disruption factors are summed for each source over the specified time period. The result is a cumulative disruption factor for each source, which indicates the relative degree to which each source is disrupting the planned flow of materials in the plant.

Chapter 7   Focusing the Improvement Process

The 80–20 rule can now be applied to prioritizing process improvements. Productivity improvement efforts should be concentrated on those sources with the highest cumulative disruption factors, since improvements in these areas will result in the greatest benefit. Managers must be careful not to avoid tackling difficult problems in favor of tackling easier problems, which have much lower cumulative disruption factors. The only valid reason to solve easier, less significant problems first is to build confidence and establish a pattern of success. Of course, the problem with tackling the smaller problems first is that they have minimal effect on improving the process.

Ongoing Improvement   Once the sources of the disruptions have been identified, the causes of the disruptions at each source must be identified. There are many possible causes of disruptions. Some of the more common causes are poor quality control procedures, inadequate maintenance resulting in frequent equipment failures, unnecessarily long setups, unreliable vendors, and poor scheduling. There are already many powerful techniques available that provide excellent guidance in solving these and other types of problems. So we will not belabor the point here.

As improvements occur, the major holes in the buffer will be eliminated or reduced. As the time buffer content profile percentages increase, the throughput of the system is increasingly protected. This has direct positive effects on the competitive advantage factors of delivery performance and product cost. Customer service clearly should be enhanced by improved on-time deliveries as disruptions to throughput decrease. Product cost is also enhanced because the smooth flow of products reduces the need for costly expediting measures.

But as process improvements continue and disruptions to the buffer decrease, an interesting situation develops. At some point, the degree of protection provided by the buffers becomes excessive. When this occurs, the time buffers can be reduced. Reductions in the time buffers result in decreased levels of inventory in the system, shorter lead times, and reduced levels of operating expense. As a result, the competitive advantage factors—product quality, delivery performance, and product cost—will all improve. The firm's competitive advantage is increased.

As a firm becomes more competitive, the market generally responds with an increasing level of demand that should lead to higher throughput for the firm. This additional throughput should be very profitable since it is not accompanied by increased levels of inventory and operating expense. The firm becomes more profitable.

The additional throughput required to meet the higher levels of demand may strain the resources of the plant. If bottlenecks develop, our attention must focus on those resources in order to produce the desired throughput. But the excess capacity of the nonbottleneck resources will also decrease as throughput increases. This means that there is less excess capacity available to overcome disruptions in the process. Previously minor disruptions may begin to cause major problems to the planned product flow, resulting in significant holes in the buffers that must be addressed.

Our process of improvement in the plant has now completed a full cycle. [5, pp. 134–139] Once again, the holes in the buffer must be analyzed, and the new sources of disruptions must be targeted for improvement. The effort to eliminate disruptions, reduce buffers, improve competitiveness, and increase bottleneck capacity and throughput is a never-ending process. New problems will continue to surface, but the firm now operates at increasingly higher levels of performance and profitability.

SUMMARY

All manufacturing organizations can be classified in one of two categories. Production-constrained firms are those organizations that cannot currently produce as much product as they can sell. Market-constrained firms are those companies that can currently produce more product than they can sell.

Production-constrained firms should concentrate on obtaining more throughput. This process must include the identification and proper management of capacity constraints in order to increase processing capacity at these critical resources. Techniques for increasing the product flow at capacity constraint resources include the elimination of idle time, the reduction of setup and processing time per unit, the improvement of quality control, the reduction of the physical workload, and the purchase of additional capacity.

A market-constrained firm should concentrate on enhancing its competitive position in the market by improving product quality and customer service, and reducing product costs. The technique of buffer management— monitoring and controlling the flow of material into the time buffers of the DBR system—is the recommended approach to improving competitiveness. Effective analysis of the time buffers is the key to identifying the chief problem areas that restrict plant performance.

QUESTIONS

1. Distinguish between market-constrained firms and production-constrained firms.
2. Discuss the different strategies that should be pursued by market-constrained firms.
3. Which synchronous manufacturing principle must be considered when performing a cost/benefit analysis at bottleneck resources in production-constrained firms?
4. Explain the concept of time buffer content profile. What is its significance?

PROBLEMS

1. Company XYZ, which has annual sales of $100 million, is a production-constrained firm. Resource R35 is *the* bottleneck resource. The labor involved in the operation of this resource costs $20 per hour. A review of the labor tickets for the past year shows that R35 was processing material an average of 20 hours a day even though it was manned for the full 24 hours. What is the dollar value of this lost time? (Assume 350 working days per year.)

2. Resource R7 has a 3 day time buffer. The first table shows the schedule of work that is currently due in the buffer. The second table describes the material that is actually present in the queue. Construct a time buffer content profile for this data and discuss the meaning of the buffer content.

Work Order	Part	Scheduled Quantity	Scheduled Hours
101	A	600	6
126	B	250	4
128	C	400	8
143	A	1200	12
148	D	200	8
150	B	250	4
156	A	600	6

Work Order	Part	Quantity Available	Hours of Work
101	A	600	6
128	C	400	8
143	A	600	6
150	B	250	4
162	E	480	4

3. This routing sheet represents the process used to manufacture product A1 at the XYZ Company. Current production for product A1 is 600 units per week (30,000 units per year) and the selling price is $80 per unit. Material used in one unit of A1 costs $20. Labor costs at the five operations vary. Labor cost is $15 per hour at operations 10, 30, and 50. Labor cost is $18 per hour at operation 20, and $20 per hour at operation 40.

Operation #	Resource	Processing Time (Minutes/Piece)	Setup Time (Hours)	Yield Percentage
10	R1	10	0	90
20	R2	12	0	100
30	R3	4	2	100
40	R4	9	0	95
50	R5	6	2	100

The XYZ Company is considering a setup reduction program to cut setup times at resource R3 in half. The program is estimated to cost $20,000 and the current costs charged to this setup average $30,000 per year.

    a. Should XYZ Company proceed with this program if it is currently production-constrained?

    b. Should XYZ Company proceed with this program if it is currently market-constrained?

4. The XYZ Company described in problem 3 is considering a quality improvement program at operation 10. The program could improve the yield at operation 10 from 90% to 95%, at an estimated cost of $50,000.

    a. Should the XYZ Company proceed with the program at either operation if it is currently production-constrained?

    b. Should the XYZ Company proceed with the program at either operation if it is currently market-constrained?

5. The XYZ Company described in problem 3 is considering a quality improvement program at operation 40. The program could improve the yield at operation 40 from 95% to 97%, at an estimated cost of $50,000.

    a. Should the XYZ Company proceed with the program as if it is currently production-constrained?

    b. Should the XYZ Company proceed with the program as if it is currently market-constrained?

6. A proposal at the XYZ company described in problem 3 recommends replacing the conventional machine at resource R3 with a newer machine. The new machine can process material twice as fast, but the more complex setups would take about 30% longer. The estimated cost of the project is $75,000.

    a. Should the XYZ Company accept the proposal if it is currently production-constrained?

    b. Should the XYZ Company accept the proposal if it is currently market-constrained?

7. At the XYZ Company described in problem 3, a cross training program has been proposed that would offload some work from resource R2 to resource R4. One minute of work per piece can be offloaded at a training cost of $20,000.

    a. Should the XYZ Company accept the proposal if it is currently production-constrained?

    b. Should the XYZ Company accept the proposal if it is currently market-constrained?

8. At the XYZ Company described in problem 3, a cross training program has been proposed that would offload some work from resource R4 to resource R2. One minute of work per piece can be offloaded at a training cost of $20,000.

    a. Should the XYZ Company accept the proposal if it is currently production-constrained?

b. Should the XYZ Company accept the proposal if it is currently market-constrained?

9. Rank in order by expected dollar benefit the proposals described in problems 3 through 8 for the following cases:

   a. XYZ Company is currently production-constrained.

   b. XYZ Company is currently market-constrained.

8

Classifying and Analyzing Manufacturing Operations

UNDERSTANDING THE PRODUCT FLOW

Manufacturing executives are becoming increasingly aware of the critical importance of the flow of material through the plant. The focus is shifting to improved product flows being the key to increased market responsiveness and reliability of delivery promises. Managers are also focusing attention on the way each product moves through the various resources (e.g., work cells and factory layouts) and on the number and diversity of products that utilize a particular resource or a series of resources (e.g., focused factories). Where a dedicated and well-informed management team leads the way, efforts to improve the product flow usually have a positive impact on the plant's performance.

Reconsidering Resource/ Product Interactions

Many companies have redesigned their plants to improve the product flow, but the actual results typically fall far short of expectations. The key reason for the unfulfilled expectations is that management's understanding of production flow has traditionally been deficient in two main areas: [4, p. 1]

1. The existence of, and the fundamental difference between, constraints and nonconstraints in a plant.
2. The interplay between current management policies and practices, including performance measurement, and the specific nature of the resources and products in the plant.

The concept of constraints has already been thoroughly discussed in Chapter 4. Significantly, there is a growing recognition of the critical importance of the basic role played by constraints in the manufacturing environment. However, there is currently no evidence to indicate that there is a general understanding of the complex interactions between current management policies and practices, resources, and product flows in the plant. As a result, there are three primary objectives in this chapter:

1. To demonstrate that the types of problems experienced by a plant can be traced to the particular and specific nature of the interactions (between resources and products) that exist in the plant.
2. To demonstrate that traditional management strategies typically do not solve, and often aggravate, the problems that characterize different manufacturing environments.
3. To identify the appropriate management policies that will optimize the performance of the plant. Of course, the appropriate policies will be a function of the type of resource and product interactions that exist in the plant.

To understand the specific problems of a given operation, the nature of the interactions in that operation must be understood. An interaction between resources is created, most simply, by the flow of product from one resource to another. For example, the engineering routing for the production of most products calls for processing to occur sequentially at a number of different work centers or resources. Clearly, the resources involved in the routing of any product interact with each other. But interactions can also be created without the direct flow of material from one resource to another. In essence, the way in which materials flow through a manufacturing facility largely determines the complex pattern of resource and product interactions that occur within the plant.

Interactions between Multiple Products and Resources

The five basic resource interactions that exist in manufacturing environments were presented in Chapter 3 to help demonstrate some of the basic principles of synchronous manufacturing. Realistically, the flow of material in a manufacturing environment is significantly more complex than these five basic cases would indicate. The product flow in most plants generally involves more than two resources, and the various resources normally process more than one product. To fully understand the effects of the numerous complex resource/product interactions that exist in all manufacturing plants, a more advanced analysis is necessary. [4, pp. 12–21] Such an in-depth analysis is beyond the scope of this book, but it will be included in the second volume on synchronous manufacturing. Nevertheless, it is possible to summarize some of the primary conclusions from such an analysis and apply the principles to the management of various types of manufacturing plants.

Managing manufacturing plants according to the traditional guidelines of efficiency and activation causes two major problems: (1) the overactivation or misallocation of resources, and (2) the misallocation of material.

Caution should be taken to distinguish between activation and overactivation. The term activation includes utilization and is not just restricted to the excess or overactivation activities.

It is possible to overactivate a resource without misallocating that resource away from the production of scheduled products. For example, if a resource has excess capacity, it can process all of the orders to meet required throughput levels and still be able to process additional excess materials. Of course, such overactivation of a resource always creates excess inventory somewhere in the system. This additional inventory may exist as either work in process or finished goods.

The misallocation of a resource automatically implies that the resource is overactivated and therefore leads to excess inventory. But resource misallocation also implies that the overactivation of a resource on one product is performed at the expense of other products. This occurs when an excessive amount of one material or product is processed while other materials needed to maintain the required throughput are left waiting in queues. Resource misallocation can cause resources to become temporary bottlenecks, resulting in production delays that lead to missed due dates. Furthermore, if the misallocation results in the starvation of a true bottleneck, then the system will lose throughput.

Material may be misallocated when a resource has the option of transforming a given piece of material into two or more different materials that are not interchangeable. The misallocation actually occurs when a resource is overactivated and processes material into a form that cannot be converted into immediate throughput.

Even though resource misallocation is harmful, the misallocation of material may have even more severe consequences. This is because a misallocation of material necessarily requires a misallocation of the resource performing the task. Also, since the misallocated material must be replaced, an additional load is placed on all resources that have processed the material up to the point of misallocation. Therefore, in managing the flow of material through a plant, particular emphasis must be given to reducing the possibility of misallocating material.

The Product Flow Diagram

The analysis of manufacturing operations begins with the development of a format that allows an accurate representation of the product flow. The general configuration, which describes the basic flow and the interactions that occur in the manufacturing process, is called the *product flow diagram.* [36]

The product flow diagram is the basic tool required for understanding production planning and control problems in any manufacturing operation. The product flow diagram provides management with a complete map of the various resource/product interactions that occur throughout the plant. Understanding the implications of these resource/product interactions allows managers to better identify the critical management problems.

The most basic element of manufacturing is the operation (or processing) performed on a specific part at a specific resource. The term *operation* is typically used in engineering to signify a process (e.g., the drilling operation). But the same operation can be performed on two different part numbers (e.g., when the only difference is the length of the material). Therefore, a new term, used to define an operation performed on a specific part at a specific resource, is needed. This basic element is called a *station.* The general information required to define a station includes part identification, the specific process identification, and the specific resource (work center) identification. To facilitate ease of understanding, in a product flow diagram the basic information that defines a specific station is generally presented in the format indicated in Figure 8.1.

FIGURE 8.1   THE FORMAT USED TO PRESENT STATION INFORMATION

Part Identification – Process Identification
Resource Identification

For example, consider the routing for the manufacture of a product identified by the part number H2786, as shown in Table 8.1. Each operation in the routing is represented by a station. Suppose that operation 030 represents a drilling operation. The corresponding station for operation 030 is characterized by the following elements:

- Part identification = H2786
- Specific process identification = operation number 030

TABLE 8.1 ENGINEERING ROUTING FOR PART H2786

PART H2786			
Operation	Work Center	Processing Time per Unit (Minutes)	Setup Time (Hours)
010	02-176	2.76	0.00
020	04-110	9.60	0.42
030	03-201	6.30	0.75
040	03-201	11.74	0.50
050	06-413	4.16	0.00
060	02-100	29.22	0.25
999	99-000	0.00	0.00

Note: The first station in this routing is defined as operation 010 on part H2786 at work center 02-176.

- Specific resource identification = drill press with the shop identification number 03-201

This station is represented graphically in Figure 8.2(a). Note that a second operation (040) performed on the same part (H2786), using the same resource (03-201), is represented by a different station. This situation is illustrated in Figure 8.2(b). Similarly, the same drilling operation (030) performed on a different part (e.g., one identified as part number G3786) would be represented by a different station. This situation is represented in Figure 8.2(c).

Each individual station identifies a specific stage in the production of a uniquely identified part. Every product manufactured by the plant will move through several resources during processing. Thus, the transformation of each product from raw material to a completed product can be represented by a unique sequence of stations. The manufacture of part H2786, as defined in the routing of Table 8.1, can be represented by the sequence of stations shown in Figure 8.3.

Arrows are used in the product flow diagram to designate the direction of the product flow. An incoming arrow at any station indicates the existence of an earlier activity in the routing that must be completed before the current station can be activated. Similarly, an outgoing arrow from a station typically identifies the next step to be performed in the routing.

A station may have either zero, one, or more than one incoming arrow. Stations with no incoming arrows represent gateway operations and indicate the entry of material into the production process. No preceding processing activity exists for these stations. All other stations will have at least one preceding station connected by one or more incoming arrows. An assembly station, of course, would require the completion of two or more preceding

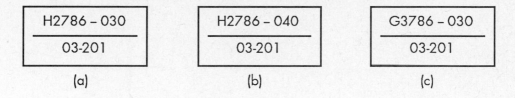

(a) (b) (c)

operations. All of the component parts required for the assembly must be available prior to assembly. Thus, a station representing an assembly operation would have multiple incoming arrows, one from each station identifying the last processing step for each required component. Figure 8.4(a) illustrates a station representing an assembly operation requiring two component parts.

A station may also have either zero, one, or more than one outgoing arrow. A station that has no outgoing arrows clearly signifies the end of the manufacturing process and the last step to be realized before shipment or sale of the product. (However, for convenience, an arrow may be used to connect the last operation to the end product.) All other stations must have at least one outgoing arrow leading to the next activity in the production process. As with incoming arrows, a station may have more than one outgoing arrow, as illustrated in Figure 8.4(b). When this situation occurs, each outgoing arrow represents a possible next activity for this product. This situation commonly occurs in industries that process a basic material into a variety of unique products. Multiple outgoing arrows also commonly identify stations that represent manufactured components used in several different assemblies.

We have identified three different types of stations:

1. Stations with no incoming arrows. These stations represent gateway operations where raw materials and purchased parts enter the process.
2. Stations with both incoming and outgoing arrows. These stations represent the manufacturing and assembly activities performed by the plant.
3. Stations with no outgoing arrows. These stations represent the final operation in the process, not including shipment of the product to a specific customer or warehouse.

These three types of stations can be combined to represent any manufacturing process. Furthermore, every product produced by the plant will involve at least one station of each type.

In addition to the first type of station we identified, a triangle may be used to represent the entry of materials into the process. When this device is used prior to a gateway operation, an arrow is typically used to connect the triangle with the gateway operation. Moreover, a triangle must be used

FIGURE 8.3   STATIONS IN THE ROUTING FOR PART H2786

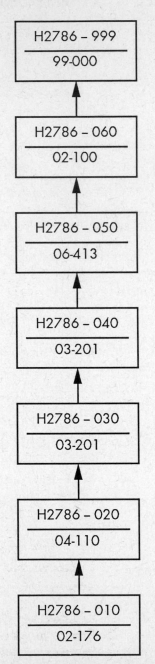

FIGURE 8.4   MULTIPLE INCOMING AND OUTGOING ARROWS

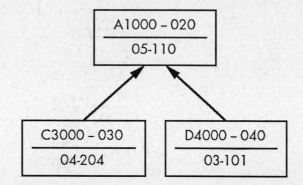

(a) Assembly Point with Multiple Incoming Arrows - Part A1000

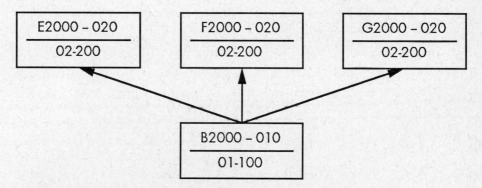

(b) Divergence Point with Multiple Outgoing Arrows - Part B2000

to indicate the introduction of additional materials or parts at a non-gateway station.

The manufacturing process for any product or group of products can be represented by a network of stations (and triangles). The resulting product flow diagram indicates the required raw materials, all of the different processing and fabricating activities performed on all of the various component parts, and all of the subassembly and assembly activities. Clearly, the product flow diagram for some manufacturing operations may be quite complex, involving a very large number of stations.

Product flow diagrams can be used to identify the dominant resource/ product interactions occurring within the manufacturing plant. The dominant resource/product interactions that characterize a plant are the key to understanding the fundamental problems that exist within that plant. Thus, the product flow diagram is a valuable tool, providing a solid basis for the application of synchronous manufacturing concepts in the plant.

CLASSIFICATION OF
MANUFACTURING OPERATIONS

Manufacturing plants may be identified as basic producers, converters, fabricators, and assemblers. A basic producer uses natural resources as raw materials and refines or separates them into products used as material inputs by converters. Converters put the materials supplied by basic producers through divergent processes to produce consumer goods or end items that can be used by a larger variety of manufacturers known as fabricators. The fabricators produce either consumer goods or products for the class of manufacturers known as assemblers. Finally, the assemblers combine various components to produce finished goods for consumers.

The product flow diagram for most complex manufacturing operations will contain a variety of resource/product interactions. However, some of these interactions will dominate the behavior of the entire operation. Moreover, all plants that have the same type of dominant interactions will have similar characteristics and problems. This provides the opportunity for the development of a general theory of synchronous manufacturing management and control that can be applied in all plants with similar characteristics and problems.

The dominant resource/product interactions that characterize different manufacturing environments provide the basis for classifying manufacturing plants into three major categories. These three categories are identified as *V-plants*, *A-plants*, and *T-plants*. Many plants fall into one of these three categories. Plants that exhibit characteristics of more than one of the three categories are referred to as *combination plants*.

The rest of this chapter describes the characteristics, consequences of traditional management practices, recommended synchronous manufacturing strategies, and illustrative case studies for V-, A-, T-, and combination plants. As this chapter unfolds, it is important to realize that manufacturing plants will evolve over time as a synchronous manufacturing environment develops. As a result, the particular set of problems facing management will change over time as the plant becomes more and more synchronized. As one set of problems is dispatched, a different set of problems will appear, but the plant will now be operating at a higher level of performance. However, it is not possible to discuss all of the possible problems that may arise as a specific category of plant evolves into a synchronous operation. Therefore, for the remainder of the chapter, we assume that all of the plant environments discussed are operating under traditional management practices and have not yet begun the evolution into becoming a synchronous manufacturing operation.

V-Plants

The class of manufacturing plants known as V-plants consists of those plants that are basic producers, converters, and fabricators. Those plants that are characterized by assembly processes are not V-plants, but are instead classified as either A-plants, T-plants, or combination plants.

Dominant Product Flow Characteristics of V-Plants   V-plants are dominated by the resource/product interaction where a single product at one stage of processing can be transformed into several distinct products at the next stage. Such a point in the product flow is referred to as a divergence point, since at this stage the flow of material diverges in several alternate directions. The product flow diagram for a plant exhibiting this basic divergence characteristic throughout the process is shown in Figure 8.5. Notice that the product flow diagram resembles the letter *V*, hence the name *V-plant*. In addition, the different products share common resources at most stages.

FIGURE 8.5   TYPICAL PRODUCT FLOW IN A V-PLANT

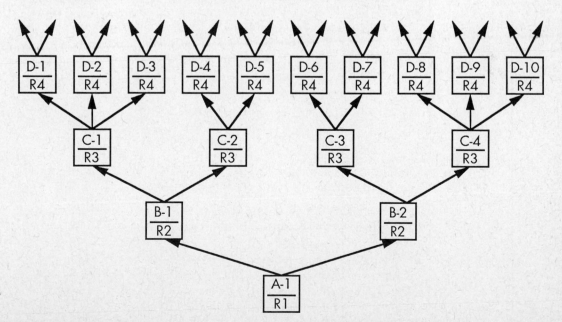

A good example of a divergent process is found in the steel industry. Consider the process where sheets of steel are rolled, hardened, and cut to exact specifications. The first stage in the process is annealing, where sheets of steel are softened in preparation for the rolling operation. At the rolling operation, a given piece of steel may be rolled into any of a large number of different thicknesses. In this process, rolling represents a divergence point in the product flow diagram. At each additional divergence point in the product flow, the number of distinctly different products continues to increase. For example, after leaving the rolling operation, the steel goes through heat treat where the material is tempered to any of a large number of combinations of desired strength and hardness characteristics. Finally, after heat treat, the steel is cut into the desired widths or strips at the slitting operation.

Figure 8.6 represents the product flow diagram for a typical steel rolling mill. The various processes involved in the operation are identified on the left-hand side of the figure. The number of different products that can be identified at each stage of the process is indicated on the right-hand side of the diagram. However, the product flow described in the V-plant diagram is generic in nature and can also be used to represent other types of manufacturing processes. For example, the same product flow diagram that describes the product flow of a steel rolling mill can also be used to describe the product flow for a textile plant. (This will be illustrated in the V-plant case study presented later in this section.)

Since the product flow diagram identifies the key characteristics of a manufacturing operation, it follows that two operations with similar product flow diagrams should exhibit similar business characteristics. It will be demonstrated that this is exactly the case. Textile mills and steel mills, for example, share many characteristics and problems. The specific characteristics and problems they share will be those associated with the existence of divergence points.

General Characteristics of V-Plants   The dominant feature of V-plants is the presence of divergence points. The existence of divergence points gives rise to three primary characteristics found in V-plants, regardless of industry type. These three characteristics are identified as follows:

1. *The number of end items is large compared to the number of raw materials.* Divergence points may exist at any stage of the production process. Hence, by the time several processing stages are completed, the number of different products being processed can be very large.
2. *All end items sold by the plant are produced in essentially the same way.* That is, all products are processed through the same basic operations and in the same sequence. For example, in the textile and steel mill

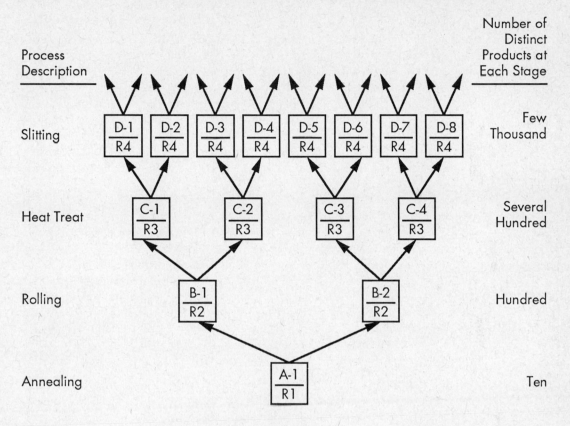

industries, every piece of finished product goes through the same sequence of operations.

3. *The equipment is generally capital intensive and highly specialized.* The evolution into this capital intensive mode of production is not difficult to understand. Since every product goes through the same sequence of operations, there are relatively few such operations in the entire plant. This contrasts greatly, for example, with a general-purpose machine shop that must be able to perform several thousand different operations. Thus, in a V-plant, it is possible to concentrate a great deal of attention on the relatively small number of basic operations that are performed repeatedly. The focus of improvement under traditional cost-based systems is to reduce the direct labor content in the product. This naturally leads to the development of specialized, high-volume, capital equipment for the basic processes involved. Direct

labor content is effectively reduced, but a typical unfortunate side effect is a loss of production flexibility.

Consequences of Traditional Management Practices in V-Plants   Perhaps the major consideration for plants that are dominated by divergence points is that each individual divergence point represents an opportunity for the misallocation of material. And when there is a sequence of divergence points in a manufacturing process, the potential for misallocation is great. This misallocation of material results in excess inventory of some products and a shortage of other products. In most cases, the excess inventory will simply inflate finished goods stocks, but sometimes it may exist as high levels of work in process. Meanwhile, the material shortage creates a constant scrambling throughout the system to expedite the materials necessary to fill customer orders.

Identifying the Problems   The types of problems encountered in a V-plant depend on whether the system has a bottleneck resource. If there is no bottleneck, then the entire plant has excess capacity. The mere existence of excess capacity, coupled with the numerous opportunities to misallocate material, is sufficient to cause significant overactivation and excessively large inventories (of the wrong products). But if the plant is also managed according to traditional cost-based performance measures, there are built-in incentives to overactivate the resources, and the magnitude of the problems is greatly increased.

If there is a bottleneck (or CCR) in the system, then the following occurs:

1. The misallocation of material and overproduction prior to the bottleneck will create a large inventory in front of the bottleneck. But because of the misallocation, this inventory is likely not to be the material necessary to meet the demand. To meet its utilization criteria, the bottleneck is forced to process the wrong products. As a result, throughput is lost and shipping schedules are jeopardized. The bottleneck also becomes subject to additional disruptions from expediters as the plant scrambles to meet shipping schedules.

2. Misallocation beyond the bottleneck has two major consequences. It results in finished goods inventories of the wrong products. It also increases the load on the bottleneck because the bottleneck has to make up for the material that has been misallocated.

Managers of V-plants are often puzzled when they have to scramble to meet the requirements of the market despite their large finished goods inventory. They invariably blame the constantly changing pattern of demand. While demand changes are natural, most of the problems are self-inflicted. The inability to respond to the market is not in spite of the inventory, but because of the inventory. The inventory was created by misallocation;

therefore, it is usually the wrong set of products. Every misallocation causes increased work loads at all upstream work centers and further constrains the ability of the system to produce the required product mix in a timely manner.

Management should understand that this misallocation is not committed in an overt and undisciplined fashion. It is not the result of malevolent managers and employees, but rather the result of carefully planned actions that focus on efficiency and cost. The misallocation is generally the result of managerial actions. Typical examples of these actions include accelerating the release of material to enhance acceptable levels of utilization and using batch sizes that are excessively large.

A major factor in the overactivation and misallocation that occurs in V-plants is the lengthy setup time often required by the V-plant resources. The long setup times are a typical by-product of the normal evolution of V-plants into capital-intensive operations. The lengthy setups encourage floor supervisors to increase batch sizes, to minimize setups by combining batches whenever possible, and to produce families of products together. These actions are consistent with traditional performance measures, but they often cause production priorities to be ignored and production lead times to become unpredictable. In addition, the large production batches cause production lead times to increase. Thus, managing according to traditional practices results in large and unpredictable lead times, which ultimately leads to missed due dates.

This discussion has focused on some of the major concerns facing the managers of V-plants. These problem areas are summarized as follows:

1. Finished goods inventories are too large.
2. Customer service is poor.
3. Manufacturing managers are uncomfortable with the apparent constant change in demand.
4. Marketing managers complain about the lack of responsiveness from the manufacturing operation.
5. Interdepartmental conflicts are common within the manufacturing operation.

Conventional Strategies for Improving Performance   In most V-plants, management's approach to improving performance centers on improving customer service and reducing production costs. The conventional approach to improving customer service normally includes increasing the level of finished goods inventory and improving the forecasting ability of the firm. The conventional approach to reducing production costs includes reducing the amount of direct labor in the product and reducing scrap and improving yields at the processes. Each of these strategies is discussed here.

Faced with large and unpredictable lead times, managers rationalize that the only way to maintain a reasonable level of customer service is to carry finished goods inventory. Thus, the excessive finished goods inventory occurs not only because of material misallocation, but also because it is planned. However, given the large number of final products typically produced by V-plants, the actual finished goods inventory usually matches poorly with market demand. This becomes a rationale for even higher levels of finished goods.

The painful realization that carrying finished goods inventories has not significantly contributed to better customer service has caused managers a great deal of frustration. The traditional solution to this problem has been to try to improve the forecasting ability of the firm. But in a competitive market, the ability to accurately forecast demand is severely limited. The producer would like firm orders as far out in time as possible, to enable better planning. But the customer prefers a short lead time for ordering material, to keep the options open. In a competitive market, the balance definitely tilts in the customer's favor. The resulting forecast accuracy in competitive markets resembles the curves shown in Figure 8.7. The fact that total sales can be predicted with some degree of accuracy is of no help to production managers. They must produce specific products. Even producing by product family is of no value. Production managers must work with the forecast for specific products or stockkeeping units (SKUs). The accuracy of this type of forecast is extremely poor; and the longer the required production lead time, the less useful the forecast becomes.

Because of the inefficiencies typically found in V-plants, and the presence of a highly competitive market, management often exerts great pressure to

FIGURE 8.7   FORECAST RELIABILITY FOR TOTAL SALES, PRODUCT FAMILIES, AND SKUs

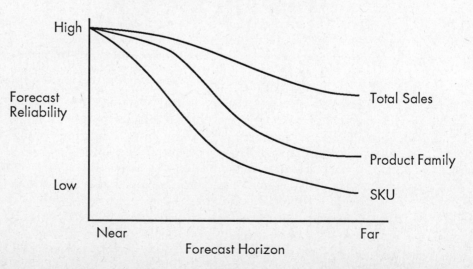

Chapter 8   Classifying and Analyzing Manufacturing Operations

reduce production costs. One traditional approach to achieving this objective is to reduce the direct labor content of the product. This will take the plant further in the direction of capital intensive operations and specialized equipment. The result is reduced labor content, but often at the expense of reduced flexibility. This may be too high a price to pay in today's competitive market.

Another conventional approach to reducing product costs is to reduce the costs incurred because of poor quality. This approach is based on the fact that quality problems result in increased costs in scrapped material, additional labor for rework, and warranty costs. The general procedure requires that the problem areas be identified and engineering solutions found. Efforts to improve quality should be applauded. But unless the improvement efforts are prioritized according to the principles of synchronous manufacturing, the systemwide impact is likely to be disappointing.

Synchronous Manufacturing Concepts Applied to V-Plants   As we have just noted, in most V-plants the ability to compete is a function of the level of customer service and product cost. The recommended synchronous manufacturing strategy to improve the performance of a V-plant is very different from the traditional approach. This strategy requires that the key competitive elements be viewed in terms of the operational measures of throughput, inventory, and operating expense.

As demonstrated in Chapter 2, the best way to improve customer service is to reduce the production lead time. Reducing production lead times is equivalent to reducing the work-in-process inventories in the system. Thus, improving service translates into reducing overall work-in-process inventories. This strategy is exactly the opposite of the traditional approach of trying to meet customer demand by increasing the amount of inventory carried.

In order to understand how to reduce the product cost, it is necessary to consider how each decision and action will affect T, I, and OE. The desired approach is to pursue those activities that increase T but reduce I and OE.

An increase in throughput requires that the firm not only produce more, but sell more. To increase sales, the competitive elements for the firm must be improved. For example, it may be necessary to improve the service level, which requires a reduction in the production lead time. To actually produce more product with the same resources requires the proper identification and management of the capacity constraints of the system. The principles of synchronous manufacturing provide the necessary guidance.

A reduction in inventory levels will immediately impact the operating expense of the business by decreasing the inventory carrying cost. But a reduction in work-in-process inventory will also positively impact the competitive elements of the business and tend to increase the throughput of the system.

Reducing operating expense will involve taking actions that will eliminate some of the wasteful elements of production, such as quality problems. In the conventional approach to improving a manufacturing operation, reductions in operating expense, as calculated by the standard cost system, are the focus of managerial actions. In the synchronous manufacturing approach, the focus is not on the calculated cost reductions, but on the global measures of T, I, and OE. Engineering and production activities will be given priority based on the estimated impact on these global measures.

The exact procedures for applying the synchronous manufacturing concepts to a specific V-plant may vary greatly from one plant to the next. These procedures, as well as the procedures for implementing synchronous manufacturing concepts in A-, T-, and combination plants, will be developed in detail in the forthcoming Volume 2 of this book. However, a brief outline of the general process for implementation in V-plants is presented at this point.

The systematic procedure begins with the identification of the physical constraints (capacity and material) that limit the performance of the system. In the case of a V-plant, there is usually only one CCR. And the location of capacity constraint resource(s) in the product flow sequence may be critical. Thus, to the extent that management can influence which of the several resources will be a CCR, this becomes an important strategic decision.

Once all capacity constraints have been identified, then the buffer requirements for the system should be determined. This involves a decision on the size and placement of both stock buffers and time buffers. Remember that these two types of buffers serve different purposes. The stock buffers are used to help service customer demand and should be located according to specific needs. The time buffers are designed to protect the system throughput from the normal disruptions that occur in a manufacturing environment. In a V-plant, time buffers should be placed only at CCRs and before shipping.

The derived master schedule for the plant must be consistent with the capabilities of the previously identified constraints. Once the master schedule has been developed, the next step is to ensure that all resources are managed to support the planned product flow. The schedule release points in a V-plant occur at material release, divergence points, and CCRs. Given the nature of the V-plant, it is likely that many work centers in this type of environment will be schedule release points.

This discussion provides a fundamental understanding of the critical issues in V-plants as well as some general procedures for managing this type of plant. Now, a V-plant case study is presented to further develop understanding of this type of manufacturing environment.

V-Plant Case Study   The A&B Company is a textile mill that produces fabrics for a variety of end uses such as upholstery and drapery products. The mill is located in the Carolinas and has about a thousand employees. The company has a long and successful history, except for the last decade. Increased competition from both foreign and domestic firms is slowly eroding A&B's

customer base. In addition, the continuing struggle for market share has created intense price competition, significantly reducing A&B's once healthy profit margins. The senior managers at A&B realize that the major competitive problem facing the company is their inability to provide satisfactory levels of customer service. In an attempt to bolster sagging profit margins, management recently established a 10 percent reduction in unit cost as a target for manufacturing managers. The customer service issue is a serious puzzle to A&B management since they have almost a 3-month supply of finished goods (the A&B inventory turns ratio is around 2.0). Thus, the challenge is to improve customer service while reducing inventories and cost.

The production process at the textile mill is quite simple. The basic production steps are yarn prep, dyeing, weaving, finishing, and sewing. At yarn prep, the yarn undergoes preparatory operations. In the dyeing operation, the properly prepared yarn is dyed to one of a large number of colors. At the weaving operation, the dyed yarn on warps is loaded onto looms where it may be woven into a number of different fabric types and patterns. At the finishing department, the fabric is fitted with backing material and sprayed with chemical finishes for durability and appearance. Finally, in the sewing department, the material is processed into the finished product.

At each step in the process, the material can be transformed into many distinct products. Figure 8.8 illustrates that a given item from yarn prep can be dyed many different colors. Similarly, yarn of a specific color from the dyeing stage can be woven into many different patterns at the looms. The product flow diagram for the A&B mill is shown in Figure 8.9. It clearly resembles the V-plant product flow diagram described earlier.

There are several significant factors that must be considered in order to manage the material flow effectively through this textile mill. Compared to a typical machine shop, the product flow appears to be relatively simple—every product passes through the same sequence of five operations. But a closer examination reveals that the flow is not as simple as it might appear.

FIGURE 8.8   THE DYEING OPERATION AS A DIVERGENCE POINT IN THE A&B TEXTILE PLANT

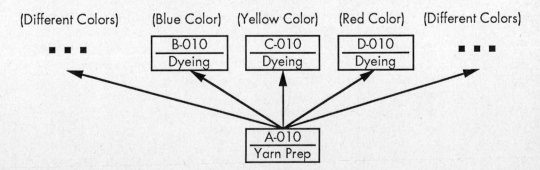

FIGURE 8.9   BASIC PRODUCT FLOW DIAGRAM FOR THE A&B TEXTILE MILL

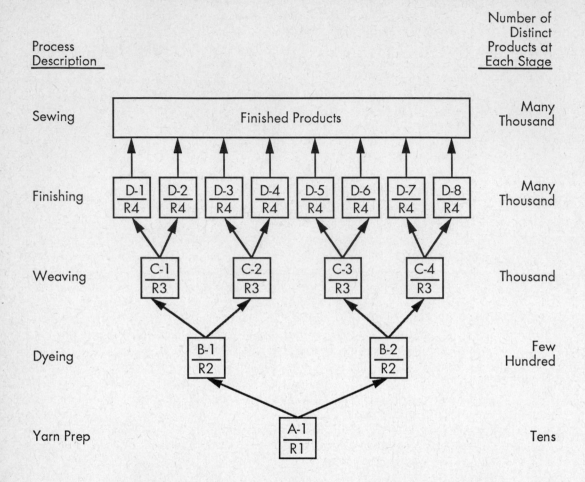

First, the number of end items is very large (thousands of distinct SKUs). Second, each end item competes for the use of the same resources (since the routings are the same). Third, although the required setup time may vary greatly from one resource to the next, the equipment is often quite complex and may require a considerable amount of setup time. Moreover, the required setup time at a given resource may vary from one setup to the next, depending on the nature of the change. For example, at the looms, changing the pattern generally requires a lengthy setup. But changing the yarn type, filler, or color at the loom can be accomplished in a much shorter period of time. A fourth factor that must be considered is that the various products exhibit a wide range of demand.

The A&B Company, like many others in the industry, operates under two basic pressures: meet the fluctuating market demand for the wide variety of products, and control the production costs of the process. As the competitive environment intensifies, the marketplace requires increased responsiveness, and management exerts more pressure to reduce production costs. As has been pointed out several times in this book, most of the traditional managerial actions that are designed to help solve one of these problems tend to aggravate the other (e.g., strategies that reduce costs tend to make production less responsive). Since production costs have been the dominant performance measure by which managers of this plant have been evaluated, customer service has generally suffered.

To clearly understand how the deterioration in customer service occurs, one must consider the cumulative effects of individual actions at each stage of production. An important consideration is the fact that the plant is characterized by expensive machinery and long setup times. Thus, the natural tendency is to process material in large batches. As a result, managers at the A&B plant established minimum production batches (or run lengths) for each of the processes. As the number of end items offered by the plant increased in response to market requirements, the quantities of each specific product ordered by customers were correspondingly reduced. The result was a mismatch between production quantities and actual market demand. This has created sizeable finished goods inventories at the plant's warehouse.

Another important consideration is that the manufacturing personnel are under continuous pressure to reduce unit costs. There are two traditional ways to achieve this objective. One way is to process materials in larger batches (by combining actual orders with other actual or forecasted orders). The other approach is to take advantage of the complex setups by processing jobs through the work centers in the sequence that minimizes total setup time. For instance, at the looms pattern changes require the longest changeovers, so production orders requiring the same pattern are grouped together. The traditional efficiency performance measures actually encourage such actions in the belief that this will make the total operation more efficient and, hence, make the plant more profitable. The fallacies of this local optimization philosophy have already been exposed in this book.

In the A&B mill, the same performance measures that are used to encourage efficiency also tend to encourage overproduction. If the loom is set up for a particular pattern, the temptation is great to continue with the same pattern as long as yarn is available or until customer pressures force a changeover. The effect of this overproduction may appear at first sight to be analogous to only an increase in the size of the process batch. However, it is far more damaging. When yarn that was prepared and dyed for one fabric is converted into another fabric, the most serious harm is that the misallocated yarn must be replaced. The overproduction at the loom, designed to save setup time and improve direct labor efficiency, has actually created a shortage of yarn required to meet an existing customer order. Overproduction (overactivation) at the loom causes both excess inventory (of the fabric produced) and a shortage (of the fabric that was supposed to be produced) at the same time.

The overactivation and consequent misallocation of material can occur at each step of the process. The result is very serious. There is a lot of inventory in the system, but significant amounts of this inventory cannot be applied to near-term customer orders. At the same time, there is a real shortage of the required fabric, causing severe customer service problems. Another significant consequence is that the amount of time material spends in the queue at any work center is highly unpredictable. Material arriving at the loom may match the pattern currently being produced. In this case, it will wait only a very short time in the queue. But if the arriving material does not match the pattern in production, it may have to wait a very long time (many weeks) before it is finally processed.

The planning lead times used for the various stages of production, indicated in Figure 8.10, clearly show the long lead times allowed for proper production sequencing. In addition, Figure 8.11 shows the typical work-in-process inventory distribution at the various stages of the process.

FIGURE 8.10   PLANNING LEAD TIMES IN USE AT THE A&B TEXTILE MILL WHEN OPERATING UNDER TRADITIONAL MANAGEMENT PRACTICES

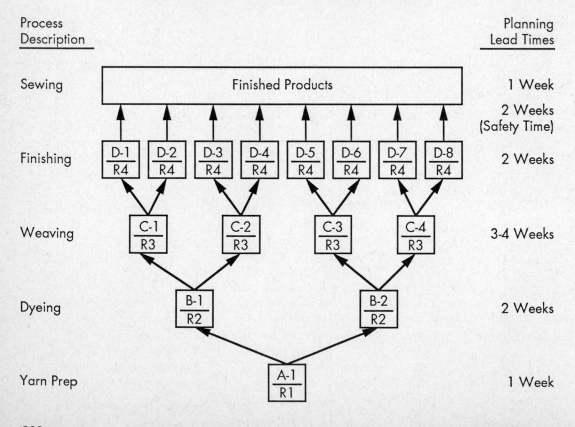

Process Description		Planning Lead Times
Sewing	Finished Products	1 Week
		2 Weeks (Safety Time)
Finishing	D-1/R4 D-2/R4 D-3/R4 D-4/R4 D-5/R4 D-6/R4 D-7/R4 D-8/R4	2 Weeks
Weaving	C-1/R3 C-2/R3 C-3/R3 C-4/R3	3-4 Weeks
Dyeing	B-1/R2 B-2/R2	2 Weeks
Yarn Prep	A-1/R1	1 Week

FIGURE 8.11   DISTRIBUTION OF INVENTORY AT DIFFERENT PRODUCTION STAGES IN THE A&B TEXTILE MILL UNDER TRADITIONAL MANAGEMENT PRACTICES

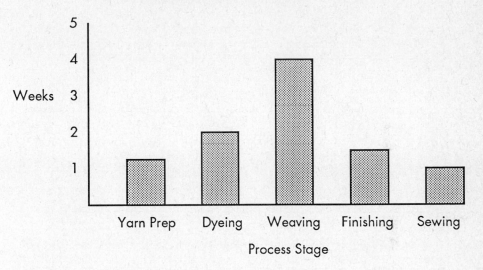

This discussion clearly shows that the problems at the A&B mill are caused by the combined effects of (1) the tendency to overactivate and misallocate that is encouraged by traditional management practices, and (2) the cumulative effect of these actions in the specific product flow characteristics of the textile mill. At the time, the managers of the A&B mill did not have the benefit of this type of analysis. As a result, management incorrectly concluded that the root cause of the poor customer service was that not enough of the right material was in inventory. Hence, they decided that the accuracy of the forecast must be improved. They made significant efforts to achieve this objective. But these efforts were to no avail, since achieving major improvements in forecasting accuracy is an almost futile task.

Concurrently, management at the A&B mill had decided to reduce the unit cost of production by reducing the direct labor content of the products. This resulted in two courses of action:

1. Replace older pieces of equipment with ones that were faster and more automated. This had been accomplished in the finishing area and in some of the warp prep areas. (But significantly, the judgment was made that it was not cost effective to replace or upgrade the looms in the weaving department.)
2. Encourage the shop floor to achieve positive labor variances by keeping busy.

Unfortunately, both of these moves did little to ease the customer service problems and, as discussed in Chapter 2, ended up increasing the overall cost of the product.

The overall approach to improving the performance of the A&B mill using synchronous manufacturing concepts is now developed. The objective is to improve customer service while reducing product costs. The most effective way to improve customer service is to reduce the production lead time. The diagram in Figure 8.12 shows the forecast accuracy for this mill. To achieve an improvement of 10 percent in the accuracy of the forecast (by SKU) was considered almost impossible. Yet a reduction in production lead time of 30 percent (from 12 to 8 weeks) will yield an improvement in the accuracy of the forecast by almost 50 percent. Figure 8.12 also indicates that cutting the production lead time from 12 weeks to 4 weeks would result in a very high level of forecast accuracy. It must be strongly emphasized that in today's dynamic marketplace, improved customer service can be achieved by reducing inventories, not by increasing them!

The problem of improving production costs was approached on two fronts:

1. Use improved deliveries and shorter lead times to secure more business. Producing more sellable product with the same resources will substantially reduce the unit cost of production.
2. Focus improvement procedures only on the areas that have a direct bearing on the constraints of the plant.

In the A&B mill, the setup at the looms was the most critical obstacle to achieving a higher level of responsiveness. Reducing the setup time at this step would have a greater impact on the bottom line of the mill operation than any automation program at any other operation.

FIGURE 8.12   ACCURACY OF FORECAST BY PRODUCT FAMILY AND INDIVIDUAL SKU OVER A 12 WEEK PERIOD

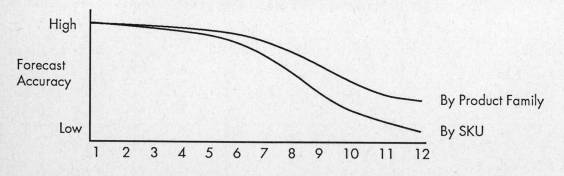

To satisfy the off-the-shelf demands of the marketplace, a stock buffer is appropriate for the finished goods area. But after applying the synchronous manufacturing philosophy and a drum-buffer-rope system to the process, the production lead time is reduced from 12 weeks to less than 4 weeks. Therefore, the finished goods inventory stock buffers can be reduced to four weeks. In addition, to protect the plant from vendor delivery problems, raw yarn was planned to be stocked in the plant. A 1 week stock buffer of raw material was deemed to be more than adequate.

Having set up the stock buffers, the raw material flow objectives were realized by effectively implementing a DBR system of material control. The looms were identified and confirmed as the CCR for the process. Rules for converting the customer demand and forecast (now restricted to a 4 week period) were developed. Essentially this consisted of setting up batch sizes (warp lengths) and sequencing rules. The batch sizes were reduced by an average of about 30 percent and sequencing rules were established that balanced due-date priorities and setups at the looms. By taking advantage of the shorter changeovers afforded by some setups, these sequencing rules enabled the planners to introduce only minimal distortion to the due-date priorities without losing capacity at the CCR. It is important to note that the objective of the sequencing rules is not to balance inventories against setup costs, but rather to balance customer priority and capacity at the CCR.

Establishing the location of the time buffers was a relatively straightforward exercise. The fact that weaving was the only CCR in the plant meant that there would be two time buffer locations. One time buffer was placed before the looms (weaving operation), and the other time buffer was established at the finished goods level. It was agreed that a 3 day buffer both at the looms and finished goods was a good starting point. Figure 8.13 shows the location and size of both the stock buffers and time buffers.

It was decided that the release of material into the system would be based on the lead times shown in Figure 8.14. However, every step in the process is a divergence point. That makes every step a schedule release point also. Thus, detailed schedules had to be released to each work center.

This procedure outlines the basic DBR approach used. Detailed rules for creating the MPS had to be developed to reflect many of the unique needs of each product and market. The mechanism for establishing the rope system and for monitoring the buffers had to be established. Most important, a step-by-step implementation plan had to be developed to help in the transition from the old planning system with a lead time of 12 weeks to the DBR system with a lead time of less than 4 weeks. The detailed procedures for implementing such a change will be discussed in the second volume of this book.

A-Plants

There are two pure categories of plants that conduct assembly operations, A-plants and T-plants. To oversimplify, those plants that build relatively few

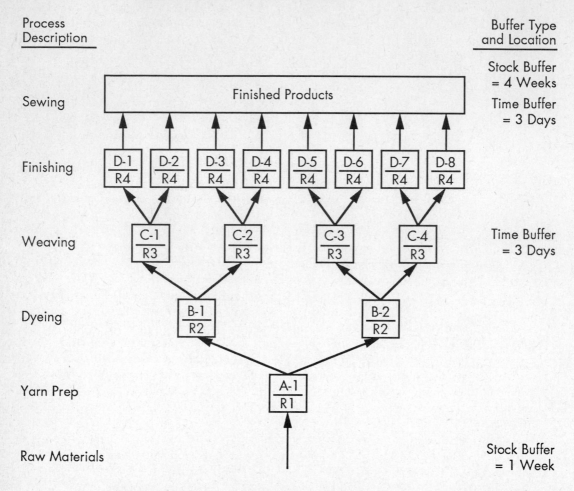

distinct products composed of mostly different components are called A-plants. For example, plants that produce heavy or specialized equipment, such as large generators, are typical A-plants.

Dominant Product Flow Characteristics of A-Plants   A-plants are dominated by the resource/product interactions where two or more component parts are assembled together to yield only one parent product. Such points in the product flow are commonly known as assembly points. But these points are also referred to as convergence points, since at these points the flow of material converges from several different sources to form a single item. A typical product flow diagram for a plant exhibiting this basic convergence characteristic throughout the process is shown in Figure 8.15. In this type of plant, in

FIGURE 8.14 LEAD TIMES AT THE A&B TEXTILE MILL AFTER IMPLEMENTATION OF THE DRUM-BUFFER-ROPE SYSTEM

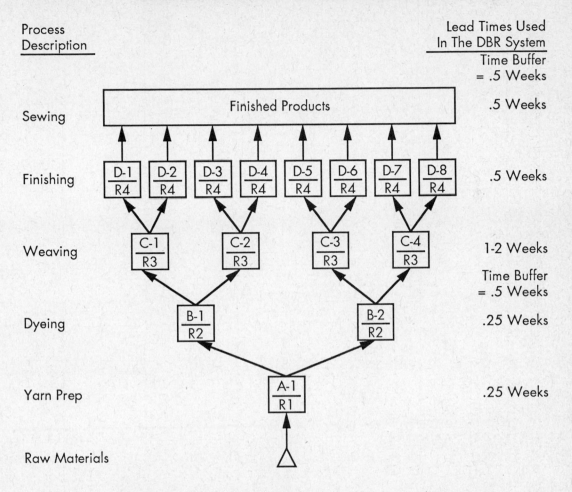

contrast to V-plants, the number of purchased materials greatly exceeds the number of end items produced. Several levels of subassemblies may be required prior to the final assembly operation. Since the overall product flow for the plant is convergent rather than divergent, the product flow diagram resembles an inverted V. Thus the name *A-plant*.

The manufacture of a jet engine is an excellent example of an assembly operation with a large number of components. To build a jet engine, many of the various components are first combined into subassemblies. In fact, most of the major component parts are subassemblies composed of several hundred parts each. The product flow diagram contains a very large number of purchased materials, all of which converge to a single product. The *A* for a jet engine assembly plant has a very wide base and a very narrow top.

FIGURE 8.15 TYPICAL PRODUCT FLOW DIAGRAM FOR A SINGLE PRODUCT IN AN A-PLANT

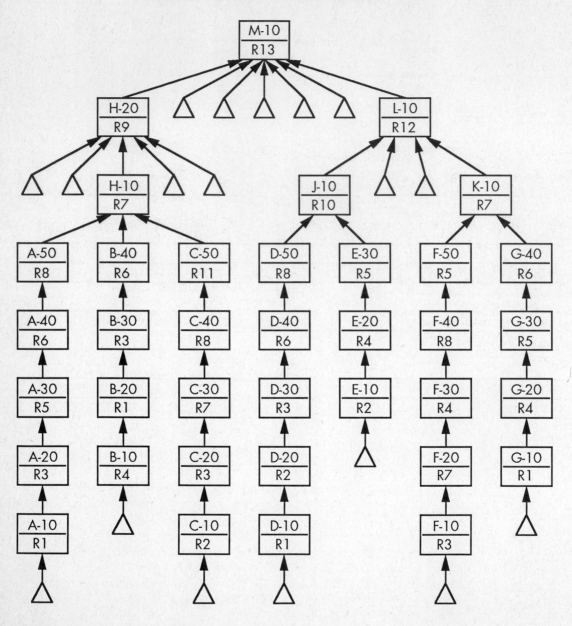

△ = Raw Material or Purchased Components

General Characteristics of A-Plants   As previously discussed, the dominant interactions represented in the product flow diagram of a firm identify the key business and operating characteristics of the plant. Thus, all firms that are dominated by the existence of convergent assembly points will share common characteristics and problems. A-plants have four dominant characteristics, identified and described as follows:

1. *The distinguishing trait is the assembly of a large number of manufactured parts into a relatively small number of end items.* Each assembly point represents a decrease in the number of distinct product types (two or more parts are combined to create one new part). After only a few assembly operations, the number of distinct products decreases dramatically.

2. *The component parts are unique to specific end items.* This is a key feature that distinguishes A-plants from T-plants. For example, consider the component parts of a jet engine. Each component part, such as a compressor blade, is unique to a specific type of jet engine. Although every jet engine will have a compressor blade, the blades are not interchangeable from one engine type to another.

3. *The production routings for the component parts are highly dissimiliar.* It is not at all uncommon that one manufactured part involves 50 operations while another part, required for the same assembly, involves very few operations.

4. *The machines and tools used in the manufacturing process tend to be general purpose.* In an A-plant, the same machine is often used to process a large number of different parts. In fact, a given part may be processed at the same machine several times in the course of its routing. Thus, much of the machinery is usually quite flexible. This contrasts greatly with the highly specialized equipment typically found in V-plants.

Consequences of Traditional Management Practices in A-Plants   A-plants are dominated by the resource/product interactions that occur at convergence points. As a result, there is usually very little opportunity for the misallocation of material in A-plants because the parts being processed are mostly unique to specific end items. Instead, A-plants are characterized by the problems that result from a misallocation of resources. In A-plants, the misallocation of resources is usually caused by products that are processed in excessively large batches. The large batch sizes are often the result of attempts to cut production costs by reducing the number of setups performed.

Identifying the Problems   Processing products in excessively large batches causes a wave-like material flow. And the use of large batches at one work center causes downstream work centers to receive material in a very erratic fashion. This erratic material flow creates two problems that appear to be contradictory but are typical of A-plants. One problem is that the utilization of resources (activation by our terminology) is not satisfactory. Because of the erratic material flow, work centers will often have to wait for material to process. This waiting causes a low level of resource utilization. The second problem is the frequent use of overtime to meet promised delivery dates. The overtime is required because the work centers are idle for long periods of time due to a lack of material. When the material finally arrives, it is too late to complete the remaining operations on time, and management must resort to the use of overtime to try to meet due dates.

Similar to V-plants, the misallocation of resources is the result of actions that are focused on cost and efficiency. It must be recognized that misallocations are not the result of a lack of discipline on the factory floor. Misallocations are the result of managerial actions such as batch sizing (to reduce costs) and accelerated release of material (to maintain labor efficiencies).

The same situation of feast or famine commonly experienced by the nonassembly work centers also exists at the assembly operations. However, unlike other work centers, assembly operations normally require that all parts be available before processing can begin. The arrival of a large batch of a single part is not sufficient to allow assembly to commence operations. The wave-like material flow makes it unlikely that all of the component parts will be available when needed. The assembly operation will constantly be short of one or more parts required to assemble the product. These missing parts must be tracked down and expedited to assembly. Because of the wave-like flow of material, the locations of the piles of inventory constantly change. This can give the false appearance of bottlenecks that wander about the plant. The resultant frequent occurrence of shortages causes constant expediting. In fact, in many A-plants, it is not stretching the truth to say that expediting is the *modus operandi*.

Managers of A-plants typically have trouble comprehending the apparent inconsistencies that plague their operation. Despite the existence of large work-in-process and component parts inventories, there is a severe shortage of parts. Interestingly enough, the misallocation of resources that creates the excessive inventories also causes the shortages and the need to expedite.

The unsatisfactory level of resource utilization and the use of overtime cause the operation to be less productive than expected. This also means that the production costs are higher than expected. The apparent contradictions between the low utilization levels and the use of overtime, and between the high levels of inventory and the constant expediting are difficult for managers to understand. These contradictions also create the impression that the operation is out of control. Managers often conclude that the production process is out of control because the information and control systems are inadequate.

This discussion has focused on some of the major concerns facing the managers of A-plants. These problem areas can be summarized as follows:

1. Assembly is continually complaining of shortages.
2. Unplanned overtime is excessive.
3. Resource utilization (not activation) is unsatisfactory.
4. Production bottlenecks seem to wander about the plant.
5. The entire operation appears to be out of control.

Conventional Strategies for Improving Performance   The traditional approach to improving the performance of an A-plant usually focuses on two issues. The strategy normally emphasizes the reduction of the unit cost of the product and the improved control of the operation.

In an attempt to reduce the product cost, management will typically stress the following:

1. *Improve the efficiency of the operation.* This effort is specifically directed at improving the efficiency of the direct labor involved in production. The low utilization seems to suggest the existence of too many production workers. Hence, in A-plants, there is often pressure to reduce the size of the labor force.
2. *Control the use of overtime.* Because of the low utilization rates, the use of overtime will be rigorously scrutinized and grudgingly approved. As a result, approval of overtime requests is often delayed. This tends to aggravate the problem, making it even more difficult to meet schedules.
3. *Focus engineering efforts on reducing the unit cost of production.* This often means replacing manual processes with automated processes. If this results in a loss of flexibility or increased setup times, then the result of automation may be to further aggravate the problem of erratic material flow.

The problem of lack of control is usually addressed by attempting to develop a single integrated production system. In large organizations, this is a mammoth task that is difficult to complete. The problem of designing and successfully implementing a single integrated system is made particularly difficult by the fact that different sections of the organization have different criteria that define a good system. Each organizational unit will lobby for a system that is designed to assist them in achieving their local goals. Since local goals often conflict with the global goal, suboptimization is the natural result. For example, the marketing/sales function is always looking for new ways to differentiate their product from the competition. This often leads to design or engineering changes. But engineering changes can cause major problems in manufacturing and are often accompanied by a loss of efficiency (at least temporarily). Thus, the local goals of marketing and manufacturing may conflict. Thoughtful deliberation is needed to develop appropriate goals and guidelines for both functional areas that maximize overall organizational goals.

Synchronous Manufacturing Concepts Applied to A-Plants   In most A-plants, the key competitive issues are a function of the major problems (poor resource utilization, excessive overtime, part shortages, and wandering bottlenecks) identified earlier. While it is likely that these problems will lead to poor customer service, the critical issue in most A-plants is that product costs are excessively high. In order for A-plants to reduce product costs, it is necessary to develop and implement a strategy that eliminates the root problems.

The recommended synchronous manufacturing strategy to improve the performance of an A-plant is very different from the traditional approach. The primary cause of the feast or famine syndrome in A-plants is the wave-like material flow. Thus, measures must be taken to eliminate or drastically reduce this phenomenon. The solution is obvious. The wave-like material flow must be replaced by a more uniform and synchronized flow. To reduce inventory levels and establish a more uniform material flow, the process and transfer batch sizes should be as small as possible. A uniform flow will enable the workload on the various assembly and nonassembly work centers to be leveled. This will largely solve the problem of low utilization and the excessive use of overtime to compensate for this lost capacity.

The drum-buffer-rope logistical system can be used to help establish the necessary synchronized material flow. The detailed procedures that exist for implementing the synchronous manufacturing concepts in an A-plant are beyond the scope of this book. But the general approach can be briefly summarized.

First, the constraints that limit the performance of the system must be identified. A-plants, unlike V-plants, are often characterized by more than one CCR. And the procedures used to identify CCRs in an A-plant are vastly different from the procedures developed for V-plants.

After the constraints have been identified, the time and stock buffers must be determined. In an A-plant, stock buffers can be used to great advantage, but in a much different way than in V-plants. Given the nature of the A-plant process, it is of little advantage to hold stocks of some component parts as work in process in anticipation of customer demand. But stock buffers can be used to simplify the problem of controlling a large number of the minor component parts required for assembly, especially those that may be common to several different end products. The time buffers in an A-plant serve the purpose of protecting the throughput from the numerous disruptions that exist in this type of manufacturing environment. In an A-plant, time buffers should be placed before the CCRs, before assembly, and before shipping.

The various system constraints must be considered when deriving the master production schedule. Once the MPS is determined, all resources in an A-plant must be managed so as to fully support the planned product flow established by the MPS. The product flow for the entire plant can be successfully managed by controlling a relatively few schedule release points. The schedule release points in an A-plant are at material release, assembly points, and CCRs. Clearly, given the large number of different resources that exist in a typical A-plant, the number of resources will normally greatly exceed the number of schedule release points.

This discussion provides a fundamental understanding of the critical issues in A-plants and outlines some general procedures for managing this type of plant. An A-plant case is now presented to develop additional understanding of this type of manufacturing environment.

A-Plant Case Study   The C&D Company is a manufacturer of pumps for the aerospace and defense industries. The plant is located in New England and employs several hundred people. Unlike many consumer and commercial product manufacturers, their business has been very solid the past few years and has actually expanded at a rate of about 15 percent per year. However, all is not roses at the C&D Company. In spite of the long lead times allowed by their customers, the plant is having difficulty shipping products within 2 weeks of the promised dates. Furthermore, expenses are alarmingly high. In fact, profits are on the verge of disappearing completely. The managers of the C&D Company recognize that they must improve their shipping performance and control expenses.

The C&D Company manufactures less than 30 end items, and each has a very specific and specialized application. Each end item is assembled from about 40 components, and the key components are produced in the plant. There is very little commonality of components between the different products. The plant manufactures an average of about 10 key components for each product. These parts, such as housings, shafts, blades, and sleeves, involve significant amounts of machine and manual work. The product flow diagram for this plant is shown in Figure 8.16 and clearly resembles the A-plant diagram shown in Figure 8.15.

The plant is organized by functional areas: the turning department, the milling department, the drilling department, the grinding department, heat treat and other batch processing departments, manual bench work, subassembly, final assembly, and testing. Compared to the A&B mill, the manufactured parts generally have long and highly variable routings, with parts having between 10 and 50 different operations in the routing. Several specialized processes are performed by outside vendors. The actual manufacturing operations are performed by general-purpose machines that work on a wide variety of products. For example, the same milling machines that process shafts also process sleeves. Most parts enter and leave a department several times (e.g., green grind to heat treat to finish grind), and the sequence varies from one part to the next.

FIGURE 8.16 PARTIAL PRODUCT FLOW DIAGRAM FOR A SINGLE PUMP IN THE C&D PUMP PLANT (EACH PUMP HAS A DIFFERENT PRODUCT FLOW DIAGRAM)

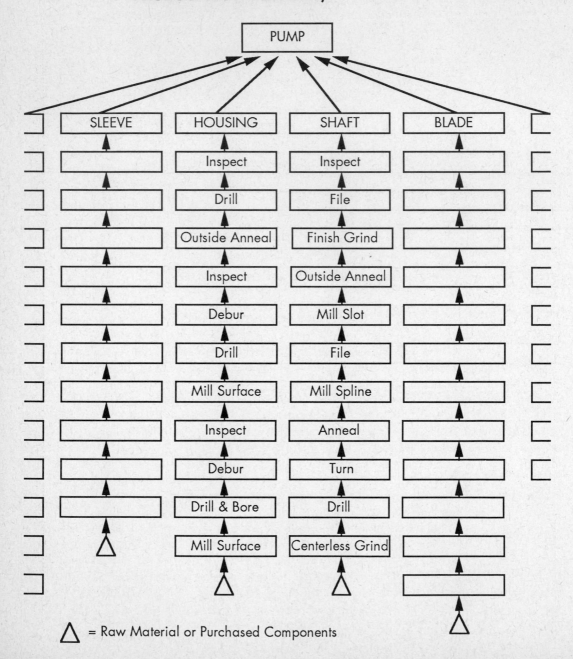

△ = Raw Material or Purchased Components

The plant manager of the C&D facility would describe the situation as follows:

> The plant did not ship the units expected to be shipped yesterday. The shafts were not available. They are still at an outside vendor and are being expedited to the extent possible. No other products were assembled because at least one required part for each scheduled product was unavailable.
>
> Despite the fact that assembly is underutilized due to parts shortages, the finance manager indicates that in-process inventories are out of control and must be reduced. A visit to the shop floor shows piles of material everywhere, most of which have a red "expedite" sticker on them. In spite of all this in-process material, end products cannot be assembled in a timely manner.
>
> Labor utilization has been running low (remember assembly did no work the previous day, just like several other work centers), but yet there are frequent requests for overtime from many departments, including assembly. The same pattern repeats every month, with the result that overtime expense is out of control.
>
> In a typical month, the plant ships only about 80 percent of the orders due that month. Furthermore, the cost of attaining even that level of delivery performance is absurdly high. Nothing ever seems to get produced without some kind of fire drill and considerable overtime.
>
> Every attempt to isolate a few simple causes for the problems has been unsuccessful. The plant appears to be plagued by every problem known to manufacturing managers: material arriving late, defective incoming material, crucial machines breaking down at critical times, excessive in-process scrap and rework, untimely absenteeism, etc.

To understand the causes for this situation at the C&D plant, it is necessary to examine the primary driving forces for this plant. Shipping the product out on time is a major consideration in the business and is part of the contractual obligation to their small customer base. And as a business enterprise, the plant is expected to show some reasonable level of profit. Customer lead time is not an issue, since the customers generally provide more than the quoted lead times. Like most such facilities, this plant tries to achieve shipping targets by allowing a generous lead time to produce the product. Management has always believed that this more than adequate lead time, coupled with a careful monitoring of schedule attainment and resource utilization, would enable the firm to meet its profit and customer delivery objectives.

It is precisely this assumption that has created the current situation in the C&D plant. To the uninitiated, it may be difficult to understand how playing it safe (by allowing plenty of lead time) and being economical (by attempting to reduce unit costs) actually precipitates the crisis. The answer actually lies in the basic nature of manufacturing operations, that is, in the cumulative effects of fluctuations and dependencies (remember the 50 operations, the common resources, the back and forth flow of materials between resources, and finally the additional dependency caused by the assembly operation).

By trying to play it safe, the managers at C&D have flooded the shop floor with material. Priorities are easily distorted as each work center tries to take advantage of the often large work queues by combining like parts, engaging in cherry picking (choosing the best or easiest jobs to work on first), and waiting until sufficient loads are available before assigning workers to operations. All of these actions occur with the intention of being economical. The amount of distortion that can be introduced is very high, and combined with the great magnitude of dependencies that exist in the plant, the material flow falls completely out of synchronization. The problem is further aggravated by the desire to keep the assembly operation active. After all, if products are not assembled, they cannot be shipped, and revenue cannot be collected.

The desire to assemble something at C&D motivates managers to evaluate what is available. If all of the necessary components except shafts are available to build a specific pump, then the expediters move into overdrive to get the required shafts to assembly. The fact that this specific pump was not due to be shipped until next week (or even next month) is secondary to being able to build and ship some product. This expediting throws all priorities aside and introduces its own significant disruptive effect into the plant's operation. Once started, the expedite function becomes the rule. The production status and component stockroom is reviewed daily to identify the parts that are worth expediting so that pumps can be assembled that day. Since the routings do not follow any specific flow, every work center at C&D is subject to the expediter's short-sighted demands. Expediting has become the *de facto* material control system in the plant.

At C&D, quality problems also severely affect the process. The reason is that when a batch of parts for one component is set aside because of quality problems, there may not be another batch of parts in the process at that time which can be easily substituted (unlike the case with repetitive manufacturing plants). In such cases, a new batch of parts must be expedited from the beginning through the long routing process and through all of the work-in-process piles.

The result of all of this economizing (batching, combining, and waiting for full loads) and expediting (to keep assembly busy and because of quality problems) is that the material flows are erratic and wave-like. Any work center can be starved for material to process one day and then be inundated with large amounts of work the next. Whenever there is a large pile of work-in-process inventory at a work station, there is a tendency to assume the station is a bottleneck resource, and overtime is used to work off the pile. This overtime is easy to justify since the work center will most likely appear to be behind schedule. The result is the wave-like flow of materials coupled with a hurry up and wait syndrome for both the operators and the batches of parts.

The problems in the C&D plant are the result of too much material (caused by excessive lead times) being processed in an uncoordinated and uncontrolled manner (due to local economizing factors). Any effective solution must directly address this issue.

There is no need for stock buffers in the C&D plant. This is because all customer orders are received with sufficient lead time to purchase all needed materials as well as manufacture and assemble the final products.

To implement the drum-buffer-rope strategy, all CCRs have to be identified. The process contains only one CCR: a computer numerical control (CNC) work center that performs a drill and bore operation in the production of housings. There is a continuous huge work-in-process queue in front of the machine, and it runs overtime on a regular basis. It is quickly determined that producing on an order-for-order basis is not possible at this machine. The load placed on the resource is close to the available capacity, so cutting the setups in half by doubling the batch sizes would do the trick. This simply means that the MPS would have to combine customer orders for 2 weeks. It is also essential to ensure that expediting does not interfere with this resource and that it is not starved for material at any time. These objectives are satisfied by treating the CNC work center as a CCR. This means that a time buffer is to be maintained at this work center, incoming parts are to be inspected for defects, and authorization is needed to run overtime at this work center to catch up to the schedule.

In addition to the time buffer placed before the CCR, additional time buffers are placed before the final assembly operation for the various final assembly components. This is done because the various disruptions throughout the manufacturing environment will randomly cause some of these components to fall behind the established schedule. The time buffers must be large enough to allow final assembly to have the full array of needed components on hand so that pumps may be assembled as scheduled. Even though the size of these time buffers may appear to be considerable, with no other planned WIP queues in the operation (except at the CCR) the total manufacturing lead times for orders is greatly reduced. Figure 8.17 illustrates the recommended location of the time buffers in the C&D plant.

Establishing the MPS is a very straightforward task in this plant. The customer orders are converted into an assembly build schedule, which loads the assembly operation to its capacity (x number of pumps per day) in the sequence specified by the order due dates and in quantities that reflect 2 weeks' worth of orders. This procedure loads the assembly operation while satisfying customer due dates. Even though the CNC machine is treated as a CCR, it does not introduce any additional modifications to the MPS because it has sufficient capacity to execute the MPS.

Finally, the rope system is established. Lead times used in the planning process are reduced to reflect the true processing times. In addition, the

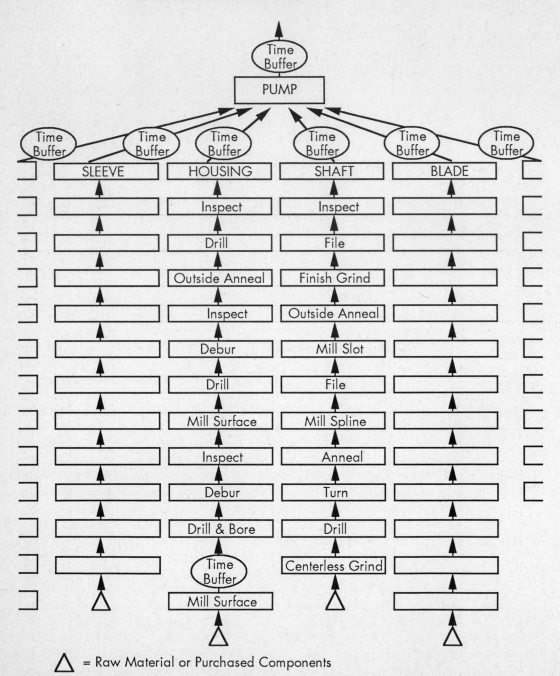

FIGURE 8.17   DIAGRAM SHOWING THE LOCATION OF TIME BUFFERS IN THE C&D PUMP PLANT

△ = Raw Material or Purchased Components

batch sizes are reduced from monthly quantities to the MPS quantity (2 week batches). Strict control over material release is exercised so that no batches are released to the production floor before the planned release date. This also implies that the assembly operation is not allowed to expedite parts for any pump that is not on the schedule for the current week.

The major cultural change, and hence obstacle, to the implementation process is the rope system. Detailed procedures for refining the MPS to meet specific constraints, and also for developing implementation plans that recognize the full impact of the DBR system, will be developed in the second volume of this book.

T-Plants

The critical feature of T-plants is that final products are assembled using a number of component parts, most of which are common to many final products. This situation usually occurs in companies that produce product families which are either highly optioned or offer a number of different packaging variations. In addition, most household and small appliance manufacturers are T-plants.

Dominant Product Flow Characteristics of T-Plants   In T-plants, a relatively small number of component parts may be combined to form a large number of different assemblies. As a result, the number of end items can greatly exceed the number of component parts. The product flow diagram thus expands at the top, resembling the letter *T*. Hence the name *T-plant*.

T-plants are typically found in an assemble-to-order environment where the required customer lead times are relatively short, component procurement and processing time is relatively long, and demand for the individual products is difficult to forecast. As a result, the components necessary to produce the various products are master scheduled and stocked prior to final assembly. The resulting interactions between the available components, required products, and limited resources dominate the T-plant environment.

The final assembly region of the T-plant product flow diagram is referred to as the top of the structure, while the region containing all of the processes prior to final assembly is referred to as the base of the structure. The top of the T-plant diagram always has the same basic structure: a set of mostly common component parts exploding into a much larger number of end items. The pure T-plant case is represented symbolically in Figure 8.18, where a number of final products (some of which are identified as E-10 through E-16) are generated from a pool of common component parts (some of which are identified as A through D). Notice that the base of the T-plant structure in Figure 8.18 contains no divergence or assembly points. A pure T-plant is defined here as one where the base structure has essentially an *I* shape. That is, purchased components are neither subassembled nor processed

FIGURE 8.18   PARTIAL PRODUCT FLOW DIAGRAM FOR A PURE T-PLANT

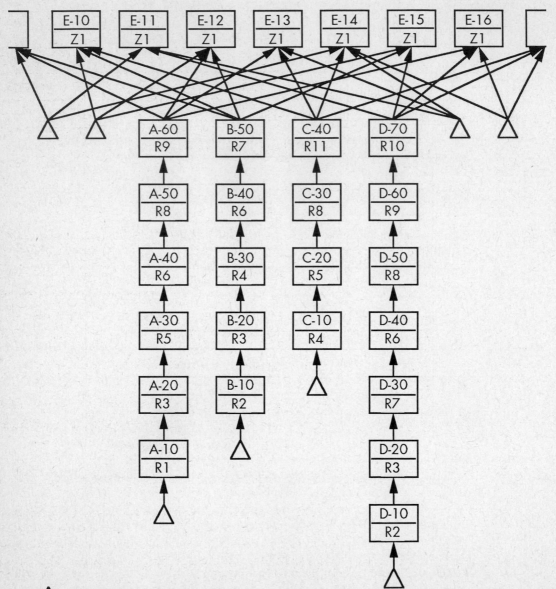

△ = Raw Material or Purchased Components

through divergence points. In this type of environment, the number of purchased components is equal to the number of components used in final assembly. This type of plant is typical of assembly plants where little or no subassembly work is performed.

A relatively simple example of a T-plant is a factory that manufactures household door locks. In order to qualify as a pure T-plant, the lock components are purchased as mostly finished parts. Any remaining processing on the component parts at the plant does not involve divergence points or subassembly. The major components of a cylindrical door lock are the inner rose, inner knob, lock body, striker plate, outer rose, and outer knob. Each of these component parts may come in a number of variations. Moreover, each variation of every component is common to many similar locks. For example, the same lock body can be used with the same style knob, but with different finishes, different style knobs for each finish, and so on. In addition, any one of the knobs used with a given lock body can be assembled with several different lock bodies. The result of this high degree of commonality is that from a finite number of knobs, lock bodies, striker plates, and roses, an almost endless variety of complete locks can be assembled. This creates the sudden explosion of the product flow diagram to create the *T* shape.

The following numerical example demonstrates the very large number of different end items that are theoretically possible from a limited number of interchangeable component parts. Suppose that any of the 6 component parts of the cylindrical door lock can be combined in any fashion with the other 5 component parts. If there are just four variations of each component part, then the number of possible end products is $4 \times 4 \times 4 \times 4 \times 4 \times 4 = 4,096$. That is, from only 24 distinctly different component parts, 4,096 end products are possible. If there were 10 variations of each component part, then there would be 60 distinct component parts, but 1 million different end items.

General Characteristics of T-Plants   At first glance, it may appear that A-plants and T-plants are very similar. It is true that they are both dominated by the interactions that occur at assembly operations, and many plants that have the distinctive T-plant characteristics at final assembly have an A-plant type of base structure. But the product flow diagram illustrates that these two types of plants have major differences. Of major significance is the fact that the nature of the assembly points for the two types of plants are exactly opposite. In A-plants, the assembly points represent an area of convergence in the product flow. But in T-plants, the assembly points represent an area of divergence in the product flow. Thus, A-plants are characterized by what we refer to as convergent assembly points, while T-plants are characterized by divergent assembly points. Additionally, the component parts used in final assembly in T-plants are common to many end items, while in A-plants the component parts are generally unique to a specific end item. These differences cause a set of problems that are unique to T-plants and require a different approach to managing and controlling the process.

It is interesting to note that T-plants and V-plants share the common characteristic of divergence. However, in T-plants, the divergence points are concentrated in one section (assembly) of the production process. In V-plants, the material flow is dominated by divergence points throughout the process; V-plants do not even have assembly points.

This discussion may be summarized by stating the four distinguishing characteristics of T-plants:

1. Several common manufactured and/or purchased component parts are assembled together to produce the final product.
2. The component parts are common to many different end items.
3. The production routings for the component parts do not include divergent or assembly processes.
4. The production routings for any component parts that require processing are usually quite dissimilar.

It should be mentioned at this point that our definition of a T-plant is somewhat restrictive. Some individuals may prefer to allow the base structure of a T-plant to include processes that are dominated by either divergence or assembly points. However, we have elected to categorize these types of plants as combination plants, which will be discussed later in this chapter.

Consequences of Traditional Management Practices in T-Plants   Since A-plants are characterized by convergent assembly points, there exists little opportunity to misallocate material. As a result, the primary problem in A-plants is the misallocation of resources. But in T-plants, the dominant interactions occur at the divergent assembly points. And because of the divergent assemblies that feature common component parts, many opportunities exist for the misallocation of materials. In fact, T-plants are dominated by the effects of the misallocation of material at the assembly operation.

Identifying the Problems   In order to demonstrate the effects of misallocation of material in a T-plant, consider the case shown in Figure 8.19. This simple case shows the assembly-level structure of the product flow diagram, which involves component parts A, B, C, and D and assembled products E, F, G, and H. The arrows between the component parts and the assemblies define how each product is made. The figure also indicates the available inventory (stock) of each component part. Now suppose that a sales order is received for 100 units of product E. The production of product E requires 100 units of part A and 100 units of part B. As seen from the diagram, although there are 100 units of part B in stock, part A is not currently available. An expediter will have to be dispatched to accelerate the arrival of part A to the assembly

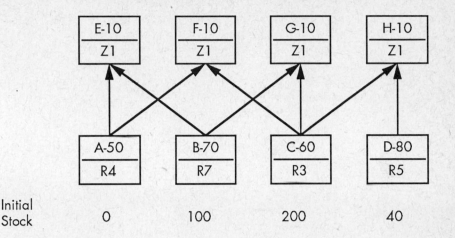

operation. (With luck, the material will be in the plant.) In the meantime, the assembly operation is idle. Although product E cannot be immediately produced, there is an adequate supply of material to produce 100 units of product G, which requires parts B and C. If the decision is made to activate the assembly operation and assemble product G, then part B and part C will have been misallocated.

The misallocation of part B to produce units of G (for which there is no current demand) leads to the inescapable effect of all material misallocation, namely inventory. Finished goods inventory of product G will have been created. But the harm of this particular misallocation is more severe than the creation of inventory. When component A is finally expedited to the assembly area, it is no longer possible to assemble product E, for which there is demand and for which part A was expedited. This is because the part B component required to assemble product E has been consumed in the production of product G. Part B, of which there was sufficient stock in the beginning, will now have to be expedited in order to process the order for product E.

Misallocation of part B has created inventory and at least a temporary loss of throughput. Furthermore, while part B is being expedited, the newly acquired stock of part A can be assembled with part C to produce units of product F. Such further activation of the assembly operation will create additional inventory and start the vicious cycle of component part shortages all over again.

The major problem of T-plants stems from the misallocation of the common component parts at the final assembly area. Another term often used to describe this specific situation in T-plants is stealing. (Manufacturing people in T-plants understand this concept well.) As a result of stealing, the inventory of finished goods is unduly high in most T-plants. Also, because of the stealing, the plant's due date performance for customer orders is unsatisfactory. In T-plants where stealing is rampant, it is common to find that the number of orders that are completed early is roughly equal to the number of orders that are completed late. In fact, in many T-plants, about 30 to 40 percent of the orders are completed early, while 30 to 40 percent are completed late. Moreover, the severity of the problems encountered is directly related to the degree of commonality of the components and the number of components required for a typical assembly.

The component manufacturing areas (which may be separate plants) must deal with the inefficiency of constant expediting. It appears to them that the managers of the assembly department have little ability to assess their component part needs. Because of the assembly department's perceived inability to adequately forecast their own requirements, the managers of the component part processing areas will attempt to compensate. In an attempt to make sure that the necessary parts are available when requested, the manufacturing managers will be generous with their forecasts. The result is that component parts production is based on an overly generous estimate of the actual usage. The overestimates and the drive to reduce the number of setups will result in batch sizes that are too large.

The large batch sizes cause lengthy production lead times. The long production lead times combined with the difficulty of predicting the demand for specific products will result in T-plants having problems similar to V-plants. The result is a large finished goods inventory (which is even larger than the generous one planned) and a poor match between what is in stock and actual demand. It is not uncommon for T-plants to have very low inventory turns (typically less than 6).

The large component part batch sizes may cause the same wave-like movement of material previously identified in A-plants. If this occurs, all of the problems associated with the wave-like flow will therefore be present in T-plants. These problems include low utilization of resources, frequent use of overtime by underutilized resources, and continuous shortages of parts at assembly (although T-plant shortages are also aggravated by stealing).

The effect of the material misallocations eventually filters back to the purchasing department. Just as the component part production schedule is disrupted by the undisciplined assembly activities, the purchase of materials and components is also adversely affected by the changing production requirements. The result is that the purchasing department will have little faith in the material forecasts based on scheduled usage. Instead, purchasing may rely on their experience to plan their material purchases. Since the actual usage figures are often greater than forecasted, purchasing will tend to inflate their planned orders. As a result, raw material stocks will tend to be too large. Thus, at all stages of the operation, from raw material stocks to finished goods, inventories are likely to be excessive.

This discussion has focused on some of the major concerns facing the managers of T-plants. These problem areas are summarized as follows:

1. Large finished goods and component part inventories.
2. Poor due date performance (30–40 percent early and 30–40 percent late).
3. Excessive fabrication lead times.
4. Unsatisfactory resource utilization in fabrication.
5. Fabrication and assembly are treated as separate plants.

Conventional Strategies for Improving Performance   The primary concern in most T-plants is the generally poor level of customer service. In this context, customer service may be viewed as either the proficiency in meeting promised shipping dates or the ability to ship off the shelf. A typical secondary concern is high product costs. Interestingly enough, these two problem areas are closely related.

A high level of customer service is difficult to achieve because of the chaos caused by stealing at assembly. And as a result, it is unlikely that the products actually produced will match scheduled production. Therefore, many customer orders are shipped late. Satisfying demand off the shelf is also difficult because having the required mix of products in inventory is highly unlikely.

Carrying higher levels of inventory is a commonly used conventional approach designed to decouple manufacturing operations (work in process) and improve deliveries off the shelf (finished goods). But this approach has proved unsuccessful. In fact, because of the adverse effects on manufacturing lead time and operating expenses, an increased level of inventory is actually detrimental to overall firm performance.

Many T-plants are found in the consumer products market. In this type of market, the price of the product is often perceived to be a major (sometimes, the primary) competitive element. Thus, in addition to trying to improve customer service, management typically makes every effort to reduce the cost and price of the product. As a result, the conventional approach to improving performance in a T-plant generally includes the following two strategies:

1. *Improve deliveries off the shelf by developing better product forecasting techniques and improving inventory planning and control functions.* In T-plants, extensive analyses of forecasting models and available and planned inventories are frequently undertaken. However, in highly competitive environments, it is usually difficult to develop accurate forecasts for specific products. In addition, due to the chaos caused by the stealing of component parts, any widespread effort to plan inventories in anticipation of forecasted future demand is doomed to fail.
2. *Reduce the product cost by improving the efficiency of the operation.* This effort may consist of attempting to improve the operation by focusing

attention on wasteful elements such as frequent setups and overtime. But consider, for example, what happens if managers try to economize on setups. As the number of setups in the system is reduced, the batch sizes are increased, and the wave-like flow of material becomes worse. This merely aggravates the problems of production lead times, inventories, and the need for overtime. The true efficiency of the plant deteriorates.

The improvement effort may also be directed at reducing the cost of the product through product design. If product design concentrates primarily on the calculated unit cost, the result may be a proliferation of products. This happens as each component is "optimally" designed to meet the requirements of the end product at the least cost. The impact this has on inventories and deliveries is devastating. The total cost for the plant is likely to actually increase due to the dramatic increase in the required number of setups.

Another avenue for improvement may emphasize the use of new technologies to replace manual processes with automated processes. However, this will be beneficial only if throughput can be increased or if labor costs can be cut sufficiently to pay for the new equipment. In addition, there must be no significant loss of flexibility due to increased setup times. Because of the forecasting issue (which does not plague most A-plants), the penalty for lost flexibility is more severe in a T-plant than in an A-plant.

Synchronous Manufacturing Concepts Applied to T-Plants The synchronous manufacturing approach to improving business performance in a T-plant is very different from the conventional approach. The primary problem faced by most T-plants is poor delivery performance. Secondary problems include excessive levels of inventory and the inability to respond quickly to a dynamic market. Two basic conditions, identified here, are essential for improved customer service and reduced inventory in T-plants.

First, the flow of product throughout the system must be synchronized. Material release, component fabrication, and assembly must be in step with demand. A critical prerequisite for achieving a synchronized flow will be the elimination of material misallocation at assembly. This will also significantly reduce the use of overtime and reduce the amount of inventory required to support the desired service level.

Second, engineering efforts must be focused on improving the operating efficiency of those elements that are most critical to the smooth flow of material to the assembly operation. This will result in a system that can be more responsive to the market. Such an effort involves looking at the entire mix of products rather than each product individually and developing marketing, engineering, manufacturing, and inventory strategies that work in unison to improve overall performance.

In order to implement the drum-buffer-rope logistical system in a T-plant, the first order of business is the identification of constraints. It should be mentioned that in many T-plants, there are no true bottlenecks. The existence of bottlenecks in a T-plant will cause major problems in the production of a large number of different end items. This scenario will usually cause management to take whatever action is necessary to increase the capacity to the level where bottlenecks are eliminated. As a result of overreacting to the threat of bottlenecks, some T-plants do not even have a true CCR. The problems experienced in meeting delivery requirements are not related to the availability of capacity, but rather to the management of capacity.

After identifying the constraints that limit the performance of the system, the buffers should be determined. The logical place to establish stock buffers in a T-plant is at the component stores level in front of final assembly, which is also the major divergence point in the flow. Many of the required minor component parts can be controlled with very little effort with these stock buffers. Time buffers in a T-plant should be placed before any CCRs that may exist, at the component stock level before assembly, and before shipping. Placing time buffers at these locations will protect the throughput of the system from the normal disruptions that occur in a T-plant.

T-plants can be best managed as two separate plants: an assembly plant that produces the end items, and a fabrication plant which supplies the component store room with the component parts. The assembly part of the operation should be run as an assemble-to-order operation, with special procedures to guard against stealing components from one order to build ahead for future orders. The nonassembly part of the plant should be run as a make-to-stock operation with the component stores functioning as its only customer. The MPS must be established to satisfy the needs of both parts of the plant. The schedule release points in a T-plant occur at material release, CCRs, and assembly.

This discussion provides a fundamental understanding of the critical issues in T-plants and suggests some procedures for managing this type of plant. A T-plant case study is now presented to develop additional understanding of this type of manufacturing environment.

T-Plant Case Study   The E&F Company is a manufacturer of industrial air tools. The company is located in the Southwest and employs about 500 people. Considering all possible variations on the tools produced at the plant, the catalog consists of tens of thousands of items. The plant was recently relocated from the North in order to take advantage of cheaper nonunion labor. However, because of the need to train new workers, the plant is only now beginning to achieve the skill levels required to produce a consistently high-quality product. But the major problem facing the E&F Company is its inability to provide timely delivery of the right products to the marketplace. This problem exists in spite of the fact that the finished goods warehouse has nearly a 3 month supply (measured in dollars) of products. All attempts to improve customer service have resulted in increased levels of inventory but have failed to solve the delivery problems. It seems that no matter what E&F has in finished goods inventory, the customers always want something else. The managers are at a loss to identify the problems at E&F. The marketing managers believe that their forecasts are as accurate as possible. They also feel that they usually allow sufficient time for manufacturing to build the products. Manufacturing managers at E&F believe that the marketing forecasts are terrible, that there is an unnecessary proliferation of products, and that there are too many small orders being accepted which the plant is not equipped to handle. They have argued that heavy investment in automation, cell technology, etc. is the only solution to the problem.

The basic problem at the E&F Company, according to top management, is the poor delivery performance. The secondary problem is the high unit cost of their products, which has severely cut into their profit margins. Naturally, the manufacturing managers feel that a smaller product line and larger order quantities would take care of both problems.

The production process at the E&F Company is somewhat typical of a T-plant. All of the various tools produced at the plant (screw drivers, nut setters, and wrenches) use a common family of motor assemblies. The motor assemblies consist of a handle housing (of which there are four types—pistol grip, in-line, angle, and push start), a motor (of which there are two types—reversible and nonreversible), and a gear mechanism (of which there are two types—single and double). Figure 8.20 illustrates part of the material requirements for some of the motor assemblies. The motor assembly is secured to the tool attachment (e.g., screw driver, nut setter) by a coupler assembly. The coupler assembly may be a positive clutch assembly, a cushion clutch

FIGURE 8.20   REPRESENTATIVE COMBINATIONS OF MOTORS, HANDLE HOUSINGS, AND GEARS FOR MOTOR ASSEMBLIES IN THE E&F INDUSTRIAL AIR TOOL PLANT

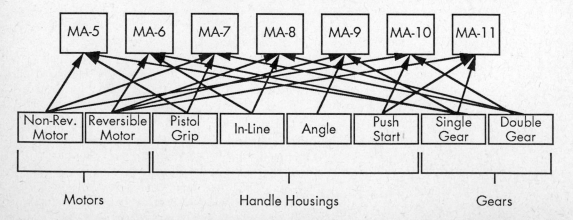

assembly, or a direct drive unit. Figure 8.21 illustrates part of the product flow diagram for the tool end items. It should be noted that both the different motor assemblies and the different coupling assemblies share many common components.

The plant manager, who is new to the plant, finds the situation at E&F thoroughly confusing. Both the component parts stock room and the finished goods warehouse are bursting at the seams. The actual total production from the assembly department seems to match the required total output. But on-time delivery performance is only averaging about 65 percent. Assembly claims that they only make products for which there is demand. But the plant manager recalls seeing stacks of obsolete products in both the component stock room and in the finished goods warehouse. Marketing design changes were blamed for these obsolete inventories. The plant manager wonders how there could be such shortages in the midst of all this inventory!

The E&F Company, like many other companies, would like to ship products from finished goods inventory. Based on the production lead time, marketing uses a fairly sophisticated forecasting module to plan the finished goods inventory for each end item. Of course, the quantity required in inventory varies from just a few units to several thousand. Manufacturing schedules replenishment products for the finished goods inventory based on the planned usage. Since component parts tend to be common across end items, replacement quantities have been established for each part using an ABC rule. Large usage items are run in 2 month quantities, medium usage items are produced in 6 month quantities, and small usage items are produced in annual quantities. The production people feel that these batch quantities are already too small to achieve really low cost efficiencies, and they view

FIGURE 8.21   REPRESENTATIVE COMBINATIONS OF MOTOR ASSEMBLIES, COUPLER ASSEMBLIES, AND TOOL ATTACHMENTS TO PRODUCE END ITEMS IN THE E&F INDUSTRIAL AIR TOOL PLANT

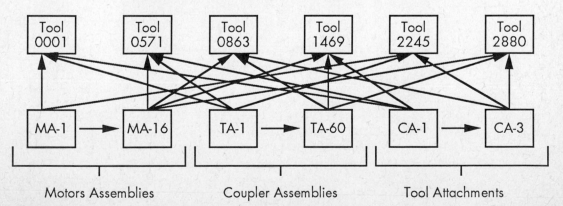

the increase in the number of medium and small usage items as a serious problem. This is partly due to the fact that the small volume parts are processed on the older conventional machines while the high volume parts are produced on the recently acquired CNC machines.

The root cause of the problems at E&F, as in the previous two case studies, lies in the inherent conflict between the needs of achieving good material flow (with high customer satisfaction) and the needs of running an efficient operation. First consider the assembly operation. At E&F, assembly is manned to achieve a daily production quantity of 2,000 units. This production target is sufficient to meet the market forecast. The output of assembly is closely tracked and has become the primary performance measure for assembly. The chaos at E&F is triggered by the fact that the focus is on total output to the exclusion of the appropriate mix of products.

There is a strong relationship between operating the assembly area according to total output quotas and the existence of activation without utilization throughout the plant. The fact that stealing occurs at E&F is evidenced by the large amount of finished goods inventory, including a significant amount of build-ahead products.

To improve the performance of E&F, the stealing phenomenon must first be controlled. To achieve this control, a strict rope system must be implemented in the plant. In addition, the performance measures should be modified. Instead of emphasizing gross output, the new performance measures should focus on customer satisfaction and producing according to the desired product mix. Producing units for which there is no firm demand must be strongly discouraged through this new set of measures.

A key factor to consider very early in the analysis is that the component parts stock room provides a better location for stock buffers than the finished goods warehouse. Establishing a stock buffer at the component part level provides many advantages:

1. The forecast accuracy at the common component level is much higher than for each possible variation of the finished tool.
2. The component part inventory is more flexible than finished goods inventory.
3. Since the lead time for assembly is less than 1 week, it is unnecessary to carry finished goods inventory for most tools. A 2 week lead time is perfectly acceptable to the market.
4. Any end item requiring less than a 1 week delivery period needs some finished goods inventory, but only enough to cover 1 week's worth of demand.

The result of establishing component part stock buffers would be that finished goods inventory can be drastically reduced. The location and recommended size of the initial stock and time buffers are shown in Figure 8.22.

The E&F facility in this case is based on a real plant that operates today as a two-plant operation in a synchronous manufacturing environment. The

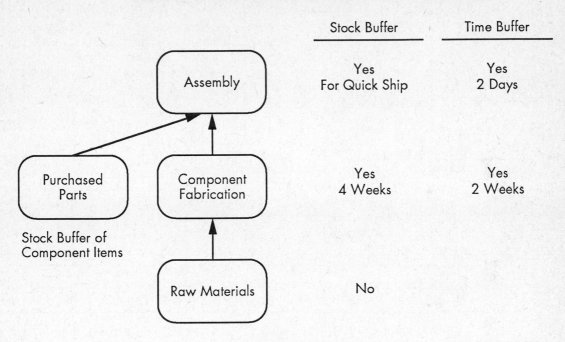

	Stock Buffer	Time Buffer
Assembly	Yes For Quick Ship	Yes 2 Days
Component Fabrication	Yes 4 Weeks	Yes 2 Weeks
Raw Materials	No	

Purchased Parts

Stock Buffer of Component Items

assembly area operates primarily in a build-to-order mode, while the fabrication areas operate in a build-to-stock mode. The fabrication areas essentially replenish the component stocks ahead of assembly. The master schedule for the assembly area is based on the weekly customer orders. For the fabrication area, the master schedule consists of the components needed for replenishment purposes. The batch quantities for each of these components is typically significantly less than the batch quantities derived from using the ABC rule.

A point of interest at the plant was that there were no CCRs in the fabrication areas. Nor were there any material constraints. All of the problems were related to managerial and behavioral constraints, not to physical constraints. Therefore, the time buffers were needed only at component stores and at finished goods. Finally, a rope system was established in the component fabrication areas to ensure that priorities would be followed. Soon after the establishment of the synchronous manufacturing concepts in the plant, lead times were drastically reduced. As a result, with less than half the inventory, due date performance improved from about 65 percent to over 90 percent. At the same time, overtime usage fell by 60 percent.

Combination Plants

Many plants fall neatly into one of the three major classifications, as did the examples used to illustrate each of the three pure cases. When a plant is clearly either a V-, A-, or T-plant, identifying the problems is a rather straightforward task. This also makes it easier to formulate an appropriate strategy to improve the operation of the plant. But there are many plants that do not exactly fit into either of the V-, A-, or T-plant categories. Such plants are called combination plants.

Plant Structures   The three pure categories of plants, V, A, and T, represent manufacturing environments where a particular type of interaction dominates the behavior of the entire plant. V-plants are dominated by the existence of divergence points throughout the production process. A-plants, on the other hand, have no divergence points and are dominated by the existence of convergent assembly operations. T-plants, like A-plants, exhibit no divergences in the production of the component parts. But the assembly operation in a T-plant represents a major divergence point that dominates the plant.

Combination plants typically occur in manufacturing facilities that have a high degree of vertical integration. In some highly vertically integrated plants, the distinctive characteristics of V-, A-, and T-plants may all be found at one facility. However, it is most common for combination plants to exhibit characteristics of only two of the three pure plant types.

There are many possible variations of combination plants. Any attempt to identify all possible combinations would likely prove futile. However, there are a few basic combinations that are sufficiently common to warrant a brief discussion. Figure 8.23 illustrates the basic structures of five different combination plants. These variations are (1) a V-base with a T-top, (2) an A-base with a T-top, (3) a V-base with an A-top, (4) a V-base with an A-middle and a T-top, and (5) an A and a V side by side topped off by a T. Each of these five variations is briefly discussed here. [37]

Five Common Combination Plants

1. *V-base with a T-top.* The base structure of this type of plant has the characteristics of a V-plant, except that the customer is another part of the plant. Typically, a few basic materials are purchased and processed through a series of divergence points to form a larger number of components or products. These items are then used at final assembly to produce a much larger number of end items. The T-plant at the end of the process is the dominant structure in this type of combination plant. The primary problems found in this type

FIGURE 8.23   THE BASIC STRUCTURE OF FIVE DIFFERENT COMBINATION PLANTS

V–Base with a T–Top

A–Base with a T–Top

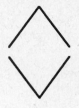

V–Base with an A–Top

V–Base with an A–Middle and a T–Top

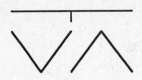
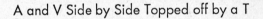

A and V Side by Side Topped off by a T

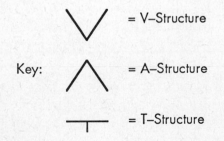

Key:

= V–Structure

= A–Structure

= T–Structure

of plant are caused by the interactions that originate at final assembly. An example of this type of plant is a vertically integrated paper mill plant, where the pulp is processed into a number of different types of paper products. The various paper products are usually master scheduled and subject to a large number of packaging variations before being shipped out to the customer.

2. *A-base with a T-top.* The base structure for this type of plant has all of the characteristics of an A-plant, except that the products at the completion of the A-process are not end items to be shipped. In this type of plant, it is typical that a large number of materials and component parts are purchased, fabricated, and subassembled to form a relatively few major assemblies that are stocked at a staging area for final assembly. These component assemblies are then combined into a large number of distinctly different end items at final assembly. Just as in the previous V/T combination, this plant is also dominated by the interactions originating at the final assembly point. The plant cannot be adequately controlled until the T-plant type problems are successfully addressed. An example of this type of plant is an assemble-to-order manufacturer of computer hardware products.

3. *V-base with an A-top.* The base structure for this type of plant has the characteristics of a V-plant except that the items produced in the V part of the process become the basic component inputs for the A part of the process. These components are then assembled (and perhaps fabricated) to form the end items for the plant. This type of plant has an interesting structure in that only a few materials are used as inputs, and there are a relatively few end items. But between the beginning and end of the process, there are a large number of different component parts. An example of this type of plant is a manufacturer of wooden chairs and tables.

4. *V-base with an A-middle and a T-top.* This type of plant is essentially a combination of two earlier cases, the V/A combination in sequence with the A/T combination. The discussion for each of these two combination plants may be applied to this case. A typical type of plant for this structure is a manufacturer of upholstered furniture. In the previous V/A example, the manufacturer produced wooden chairs and tables. Now suppose that instead of chairs and tables, the manufacturer specializes in upholstered chairs and sofas. The V/A part of the process now yields the frame for the various chairs and sofas being produced. The T part of the process is the addition of the stuffing and any of a large number of patterns and materials to the frames to produce the highly differentiated final products.

5. *A and V side by side topped off by a T.* In some cases, different processes exist side by side with neither process feeding the other. In this particular case, an A and a V exist side by side and their output becomes the input for a T-plant structure at final assembly. An example of this is a plant that produces magnetic tape in cassettes. The production of the magnetic tape is a V process, while production of the cassette container has an A structure. The tape and the containers are combined at final assembly. This part of the process has a T structure since relatively few component parts can be combined to produce a much larger number of distinctly different end items.

Synchronous Manufacturing Concepts Applied to Combination Plants   The greater the degree of vertical integration in a plant, the more complex the problems become, and the more difficult the plant is to manage and control. However, the problems found in the pure V-, A-, and T-plants will also be found in those parts of the combination plants that resemble the pure counterpart. The basic analysis for the pure V-, A-, and T-plants therefore extends to, and is valid for, the various combination plants. One approach that may simplify the management process is to use a focused factory approach to plan and control each different structural part of the plant as a separate process and as a pure case of a V-, A-, or T-plant. A more in-depth discussion and illustration of combination plants may be found in the second volume of this book.

SUMMARY

The performance exhibited by a manufacturing operation is a function of the interplay between the various resource/product interactions and the production policies enacted within the firm. To systematically expand and apply the synchronous manufacturing concepts to complex real life manufacturing operations, the full spectrum of interactions between multiple resources and products must be understood. The product flow diagram represents the configuration of interactions in the manufacturing environment and therefore is a valuable tool when analyzing operations.

The product flow diagram is very useful in describing manufacturing operations that have similar characteristics and problems. Manufacturing plants may be categorized as either V-plants, A-plants, T-plants, or some variation or combination of these three basic categories. It is now clear that companies in widely different industries, making vastly different products, exhibit many of the same product flow characteristics. Moreover, it has been demonstrated that manufacturing companies with similar product flows share similar problems.

It has become evident that the policies employed by too many manufacturing companies are based on the standard cost system. On close scrutiny, we find that these systems encourage actions that are often counterproductive to the overall productivity and profitability of the firm. The inescapable conclusion is that in order to dramatically improve our manufacturing operations, it is necessary to consider alternative ways of managing them.

The synchronous manufacturing approach to managing our organizations requires a clear recognition of the basic structure of the plant as well as the various problems that exist and their root causes. Using this understanding as a springboard, managers can apply the comprehensive and systematic set of principles, guidelines, and procedures of synchronous manufacturing to lead their organizations to increasingly higher levels of competitiveness and profitability.

QUESTIONS

1. Identify and discuss the two major misallocations that are caused by traditional production policies.
2. Which of the two misallocations identifed in question 1 is the most serious? Why?
3. Define a station. How many types of stations are there?
4. What is the significance of the product flow diagram?
5. Describe the basic differences in the material flow characteristics of V-, A-, and T-plants.
6. Discuss some of the basic problems that arise as the result of managing V-, A-, and T-plants by traditional methods.
7. Discuss the placement of stock and time buffers in a typical V-plant, a typical A-plant, and a typical T-plant.
8. Identify two specific manufacturing plants with which you are familiar. Attempt to categorize these plants as either V, A, T, or some combination of the three.

APPENDIX
The Production Dice Game

The production dice game is designed as an exercise to demonstrate the root causes of some of the problems that exist within manufacturing plants—problems such as excessive inventory, difficulty in meeting shipment dates, and the need for unplanned overtime. Most manufacturing operations share these problems because they arise from the same common causes. The production dice game is designed to enhance understanding of these root causes and assist in the establishment of some basic guidelines by which to better manage a manufacturing facility. [1, 2, pp. 175–185]

All manufacturing processes share two phenomena, irrespective of the products they produce, the processes they employ, and the markets they service. These two phenomena are dependent events and statistical fluctuations. Together they cause the performance of a manufacturing plant to be less than optimal when managed under traditional guidelines. A new approach that recognizes the combined adverse affects of these two phenomena is needed so that we may find better ways to manage our factories.

SETTING UP AND PLAYING THE GAME

The game is played by creating a simple manufacturing operation. The material processed by this simulated plant can be represented by any readily available material in the factory. For maximum effect, it is best to use materials actually processed by a plant. This will help participants relate the lessons of the game to the manufacturing environment. The only requirements for the material chosen is that it be something which can be set on a table and which people can move easily. Several hundred pieces of this material are needed to play the game. A team of six to eight players is set up to represent six to eight sequential operations required to process material (see Figure A.1). If there are more than eight players, it is better to arrange them

FIGURE A.1   SETTING UP THE PRODUCTION DICE GAME

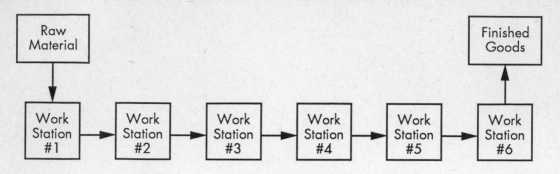

as several independent lines of five to eight people each. More complicated flows can be represented, but the basic effects of dependent events and statistical fluctuations are very well illustrated by this simple linear production line making just one product.

Each player represents one work station. Material is processed through each work station by moving it from the player's left to the player's right (or vice versa). On each day (turn), each player/work station processes pieces of work by moving them to the right. When a given piece has been processed by the last player on the right, it is considered as a unit of finished goods and counted as throughput.

The dependent events in this game are obvious. Pieces cannot move on to any work station without first passing through the previous work station.

Ordinary gaming dice are used to create the effect of fluctuations. Each player/work station is given a single die. The roll of the die is used to represent the actual production capability of each work station on a given day. A low roll of the die represents a day plagued with breakdowns, absenteeism, accidents, and similar problems. A high roll represents a good day with few problems and a high output potential for that work station. The random and unpredictable nature of the problems encountered in our plants is represented by the random nature of the outcome of the roll of the die. However, over a large number of rolls, the average outcome should be about 3.5 (since each number between 1 and 6 is equally likely). Thus, the average production capability of each work station is 3.5 units per day.

One toss of the die determines one day's worth of production at each station. After the toss, the player moves forward (processes) the number of pieces indicated by the die. If the number on the die exceeds the number of pieces available in the work-in-process queue at the beginning of the day, then only the pieces available can be moved forward that day. Do not use the pieces processed at a preceding station the same day—this is tomorrow's work queue. It is therefore possible that a station cannot produce to its capacity due to the lack of material, a common occurrence in real manufacturing plants that surfaces as lost production capacity.

Set up the game with an initial work-in-process queue of four pieces at each work station (this represents slightly more than an average day's worth of material). The first operation is the only exception to the starting WIP. The first operation should be given an adequate supply of material to cover 20 days' worth of work. See Figure A.2 for a pictorial representation.

With each operation being able to process an average of 3.5 units per day, it appears that the plant should be able to meet a market demand of 70 units per month (20 working days). Of course, the actual production will not be 3.5 units every day. There will be good days and there will be bad days. But over the course of 20 days, the plant would be expected to produce 70 units.

Only a single product is manufactured in a simple linear process, and the average capacity of each work station is exactly the same and matches the average required output rate. Therefore, this process appears to be very simple to manage with the full expectation of meeting production goals. The only problem appears to be the random production of each work station. While this may cause some temporary problems, we expect the fluctuations to average out.

But will they?

The performance of the plant and that of each work station are to be recorded on the form provided (see Figure A.3). The same form is used for each work station. Data about overall plant performance (actual versus expected units completed at the last work station) at 2 and 4 week intervals can be conveniently recorded at the bottom of the form. Data about individual

FIGURE A.2   STARTING POSITION FOR THE BASIC PRODUCTION DICE GAME

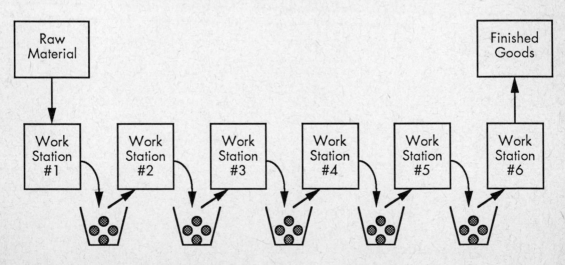

Key: ● = 1 Piece of Work-in-Process

FIGURE A.3   DICE GAME SCORE SHEET

Operational Performance For Work Station #____

(1) Roll Number (Day Number)	(2) Actual Roll	(3) Pieces Processed	(4) = (3)/(2) Efficiency Percentage	(5) WIP Pieces	(6) Overtime Rolls
1					
2					
3					
4					
5					
6					
7					
8					
9					
10					
11					
12					
13					
14					
15					
16					
17					
18					
19					
20					
Total					

Overall Plant Performance

Week Number	Expected Shipments	Actual Shipments	Past Due Shipments	Plantwide WIP	Overtime Used
1 and 2					
1,2,3, and 4					

work station performance is collected at each work station on the top part of the form. The actual roll is the number on the die for that day, representing processing capability for the day. The pieces processed column is the number of pieces actually moved to the next work station. This number cannot exceed the actual roll of the die or the number of pieces of material in the work-in-process queue for that day. The efficiency of the work station is the ratio of pieces processed to the actual roll. The work in process is the queue inventory in front of the work station at the end of each day after the player has completed moving the material and has received material from the previous work station. This should give us the information necessary to identify which operation is responsible for unsatisfactory performance.

When the results at the end of 20 days are examined, the performance of this simple process will be well below expectations. The results will be as follows:

1. Total units produced and shipped will be significantly lower than 70 units.
2. The inventory at the end of 20 days is significantly higher than the starting WIP inventory.

The inescapable phenomena of dependent events and statistical fluctuations have combined to cause this poor plant performance. At the beginning, it looked like a simple and easy-to-control plant. After 20 days, it resembles a real-life plant, i.e., one with randomly distributed piles of excess inventory that has trouble meeting shipment schedules. If the phenomena of dependent events and statistical fluctuations wreak such havoc in a simple linear process, it is easy to imagine what damage they cause in more complex real-life plants.

VARIATIONS TO THE BASIC DICE GAME

There are many variations to the basic production dice game. Your imagination is the only constraint on the possible variations. However, two instructive variations—the use of overtime and an unbalanced plant—are described below.

Whenever the production output of our plants is less than the expected or required amount, we often resort to the use of overtime. However, this is done in a reactionary fashion and does not increase shipments to the degree expected. The concept of overtime can be incorporated into the game by allowing one player an additional roll of the die after everyone has completed production for a given day. To play this variation, appoint one player as the plant manager. He is responsible for deciding who uses the available overtime. Set a limit of five to ten overtime rolls for the plant as a whole during the 20 day period. Overtime rolls are recorded in the column provided in the score sheet. Note that the work in process of the next work station (the one following the operator working overtime) will have to be updated. For each additional roll of overtime, deduct the revenue equivalent of one piece from the output. The plant manager is responsible for the total shipments, the total inventory, and the overtime used. See how well the plant now performs!

A manufacturing operation in which the available capacity at different resources is not exactly matched with the market demand, and in which the available capacity at various resources is different, is called an unbalanced plant. All real-life manufacturing operations are unbalanced.

What is the effect of dependent events and statistical fluctuations in an unbalanced plant? To find out, modify the dice game setup to represent an unbalanced plant. One easy way is to change the average processing capability of one or more stations by allowing the player at that station to use more than one die. (An interesting alternative is to substitute gaming dice that have 4, 8, 10, 12, or 20 sides. Such dice may often be found in hobby shops.) Whatever dice the players are given, the roll of the dice still represents the maximum number of units that can be processed in a given day. A recommended version is to introduce a single bottleneck into the process by giving one player less processing capacity than the other players (for example, by giving one player a 4 sided die that is numbered from 1 to 4). Start with the bottleneck station located at the end of the process (the last player). Play the game as before and see what happens. Change the location of the bottleneck in the production process and note how the location of the bottleneck station affects the process. Pay special attention to how the efficiencies of the various stations are affected by whether they are before or after the bottleneck station.

Note: Simple versions of V-, A-, and T-plants, as presented in Chapter 8, can be simulated by using dice. The production dice game can also be used to help in understanding the drum-buffer-rope concept explained in Chapter 6. The players should try to develop a DBR system that can be used to successfully manage a dice game system that has a bottleneck resource located anywhere in the process except at the gateway operation.

REFERENCES

The following books deal specifically with various aspects of synchronous manufacturing:

1. Goldratt, Eliyahu M., and Jeff Cox. *The Goal.* Revised ed. Croton-On-Hudson: North River Press, Inc., 1987.
2. Srikanth, Mokshagundam L., and Harold E. A. Cavallaro. *Regaining Competitiveness—Putting the Goal to Work.* New Haven: The Spectrum Publishing Co., Inc., 1987.
3. Srikanth, Mokshagundam L. *The Drum-Buffer-Rope System of Material Control.* New Haven: The Spectrum Publishing Co., Inc., 1987.
4. Srikanth, Mokshagundam L. *Resource Product Interactions and the Classification of Manufacturing Operations.* New Haven: The Spectrum Publishing Co., Inc., 1987.
5. Goldratt, Eliyahu M., and Robert Fox. *The Race.* Croton-On-Hudson: North River Press, Inc., 1986.

The following books provide excellent analysis of the loss of competitive position by U.S. manufacturing companies and the crucial role of traditional management and economic approaches in this decline:

6. Johnson, Thomas H., and Robert S. Kaplan. *Relevance Lost: The Rise and Fall of Management Accounting.* Boston: Harvard Business School Press, 1987.
7. Hayes, Robert H., and Steven C. Wheelwright. *Restoring Our Competitive Edge: Competing through Manufacturing.* New York: John Wiley & Sons, 1984.
8. Skinner, Wickham. *Manufacturing: The Formidable Competitive Weapon.* New York: John Wiley & Sons, 1985.
9. Piore, Michael J., and Charles F. Sabel. *The Second Industrial Divide: Prospects for Prosperity.* New York: Basic Books Inc., 1984.

The following books describe the many successful production systems from an empirical standpoint:

10. Ford, Henry. *Today and Tomorrow*. Garden City: The Garden City Publishing Company, 1926.
11. Ohno, Taiichi. *Toyota Production System: Beyond Large Scale Production*. Cambridge: Productivity Press, 1988.
12. Deming, W. Edwards. *Quality, Productivity, and Competitive Position*. Cambridge: MIT Center for Advanced Engineering Study, 1982.
13. Hall, Robert. *Zero Inventories*. Homewood: Dow Jones Irwin, 1983.
14. Schonberger, Richard J. *Japanese Manufacturing Techniques*. New York: The Free Press, 1982.
15. Schonberger, Richard J. *World Class Manufacturing*. New York: The Free Press, 1986.
16. Suzaki, Kiyoshi. *The New Manufacturing Challenge*. New York: The Free Press, 1987.
17. Shinohara, Isao. *Miracle of NPS: New Production System*. Cambridge: Productivity Press, 1988.

The following books describe the traditional approach to operations management and production planning and control:

18. Chase, Richard B., and Nicholas J. Aquilano. *Production and Operations Management—a Life Cycle Approach*. Homewood: Richard D. Irwin, Inc., 1989.
19. Wight, Oliver. *MRP II: Unlocking America's Productivity Potential*. Boston: CBI Publishing Co., Inc., 1981.
20. Orlicky, Joseph. *Material Requirements Planning*. New York: McGraw Hill, 1975.

The following are articles and documents from professional and business journals, government publications, trade publications, and conference proceedings. They are specifically referenced in the text and are listed in the order in which they first appear.

21. "Productivity Dropped 1.7% in Fourth Quarter." *Wall Street Journal* (February 3, 1987).
22. Council of Economic Advisors. *Economic Indicators*. Washington, D.C.: U.S. Government Printing Office, 1988, 36.
23. "The Hollow Corporation." *Business Week* (March 3, 1986), 56+.
24. Schmenner, Roger. "Every Factory Has a Life Cycle." *Harvard Business Review* (March/April 1983).
25. Plossl, George W., and Raymond L. Lankford. "The Redirection of U.S. Manufacturing: Part 2—The Pivotal Period." *Production and Inventory Management Review* (November 1984), 50–51.
26. Whiteside, David, and Jules Arbose. "Unsnarling Industrial Production: Why Top Management Is Starting to Care." *International Management* (March 1984), 20+.

27. Goldratt, Eliyahu M. "Cost Accounting Is Enemy Number One of Productivity." *International Conference Proceedings, American Production and Inventory Control Society* (October 1983).

28. Kaplan, Robert S. "Yesterday's Accounting Undermines Production." *Harvard Business Review* (July/August 1984), 95–101.

29. Longmire, Robert J. "Cost Accounting in a CIM Environment." *Production and Inventory Management Review* (February 1987), 47+.

30. Haft, Ronald R. "Should Machine Tools Run Faster?" *Tooling and Production* (February 1987), 59–61.

31. Goldratt, Eliyahu M. "The Unbalanced Plant." *International Conference Proceedings, American Production and Inventory Control Society* (October 1981).

32. Fox, Robert E. "OPT: An Answer for America (Part II)." *Inventories and Production* (November/December 1982), 10–19.

33. Goldratt, Eliyahu M. "100% Data Accuracy—Need or Myth?" *International Conference Proceedings, American Production and Inventory Control Society* (October 1982).

34. Fox, Robert E. "Theory of Constraints." *NAA Conference Proceedings* (September 1987).

35. Fox, Robert E. "Keys to Successful Materials Management Systems: A Contrast between Japan, Europe, and the U.S." *International Conference Proceedings, American Production and Inventory Control Society* (October 1981).

36. Maturo, Michael P. "Language of OPT." *International Conference Proceedings, American Production and Inventory Control Society* (October 1983).

37. Finch, Byron J., and James F. Cox. "WIP Inventory as an Asset: The Impact of Bill of Material Shape, Production Stage, and Plant Type on Its Strategic Use." *Production and Inventory Management* (forthcoming).

INDEX